Structure and Spectra
of Atoms

Structure and Spectra of Atoms

W. G. RICHARDS
Physical Chemical Laboratory
University of Oxford

and

P. R. SCOTT
Charterhouse, Godalming

JOHN WILEY & SONS
Chichester · New York · Sydney · Toronto

Library of Congress Cataloging in Publication Data:

Richards, William Graham
  Structure and spectra of atoms.

  Includes index.
  1. Atomic theory. 2. Atomic spectra. I. Scott,
P. R., joint author. II. Title.
QD461.R478   541'.24   75-37797

ISBN 0 471 01654 3 (Cloth)
ISBN 0 471 01759 0 (Pbk)

Typeset in IBM Century by Preface Ltd, Salisbury, Wilts
and printed in Great Britain by The Bath Press, Avon

Preface

After several years of teaching Oxford chemistry students about the structure and spectra of simple atomic systems and waiting in vain for a more gentle introduction than currently available texts, we have felt compelled to write our own.

The subject is so fundamental that it must necessarily be studied as soon as the study of chemistry is taken seriously. In the British system this means immediately on entering a University, probably during the first few weeks. In countries which have a broader high school and university education it may come later when specialization gets under way. For both categories of student it is very important to have a qualitative understanding long before it is necessary to delve into the mathematical basis of the subject or the more recondite examples.

Our approach here is always to use qualitative pictorial arguments where possible and to avoid the mathematical niceties which are to be found in many excellent advanced works. Occasionally we have even sacrificed rigour for the sake of clarity in the hope that the non-mathematical reader will still be able to understand the subject.

Despite the text being a gentle introduction, in order to emphasize the relevance of atomic spectra to chemistry, we have included references to some topics which are distinct branches of spectroscopy in their own right. These topics depend on the sets of energy levels encountered at the first look at the subject.

We wish to express our thanks to several colleagues who have helped us by reading all or part of the manuscript; especially R. F. Barrow, Susan Dean and Steven Moore.

Contents

I

Quantized Energy Levels

Over the last hundred years studies of the interactions of light and matter have played a critical role in formulating our ideas of atomic and molecular structure. These studies have ranged from the use of high-energy γ-rays to low-energy radio waves, and from the simplest atoms to giant macromolecules; in this book we shall concern ourselves solely with the interactions of infrared, visible and ultraviolet light with atoms. These interactions cause rearrangements among the outer valence electrons, and give rise to line spectra. Interpretation of these spectra will give us information about the motions of electrons in atoms; this will help us to understand how and why atoms come together to form molecules, and will explain the structure of the Periodic Table. We shall see how details of the spectra reveal that both electrons and nuclei have spin, and how these spins can be used to obtain important chemical information. A study of the effects of electric and magnetic fields will provide models for the description of the structures of molecules, in particular of transition metal complexes.

Many of the ideas that we shall meet in this book are now so familiar to us that it is difficult to appreciate how revolutionary they were; yet the work of Planck, Bohr and de Broglie at the beginning of this century led to a complete reappraisal of physics, which in turn has transformed chemistry from an art into a discipline firmly based on theoretical foundations.

A. Absorption Spectra

It is well known that if white light is passed through a diffraction grating or a prism, it is split up into its component colours, each of which corresponds to a different wavelength. If light from a hot filament lamp is analysed in this way, it is found that over a wide region of the spectrum all wavelengths of light are represented, and the light is said to be spectrally continuous. However, if the light is first passed through an atomic vapour and then analysed with a diffraction grating, it is found that the emergent light is no longer continuous. Figure 1.1 shows a simple arrangement for observing this effect. Light of different wavelengths is focussed onto different areas of the photographic plate, and in the absence of an atomic sample, the whole plate is blackened by the incident light. But when an atomic vapour is introduced into the

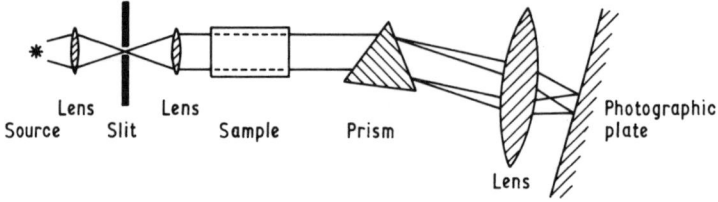

Figure 1.1 Arrangement suitable for the observation of the absorption spectrum of an atom

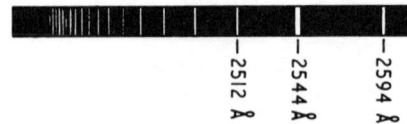

Figure 1.2 Low resolution absorption spectrum of sodium

sample chamber, it is found that light of certain wavelengths is absorbed by the vapour, and the photographic plate therefore contains light and dark regions. Figure 1.2 shows a typical plate obtained with sodium vapour; each line corresponds to a particular wavelength, and implies that that wavelength is being absorbed by the sodium vapour. This pattern which is obtained is called the absorption spectrum, and is characteristic of the atomic vapour in the sample chamber. If a different vapour were used, a different spectrum would be obtained; nevertheless it would still consist of a series of lines, and would thus differ from the absorption spectrum of a molecule, which would contain a number of bands.

B. Emission Spectra

When an atomic vapour is subjected to a high potential difference, a characteristic glow is produced. This effect has been known for many years, and is familiar to us from sodium street lamps which produce an orange-yellow light. The light emitted by such lamps can be analysed by

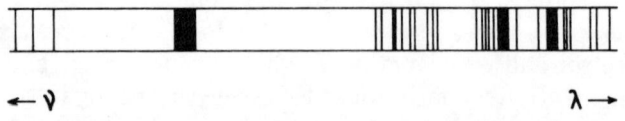

Figure 1.3 The emission spectrum of mercury

a diffraction grating; again it is found that a series of lines is produced, although in this case the lines appear dark on the photographic plate, and the background is light. Figure 1.3 shows a typical plate using a mercury lamp. This pattern of lines is called the emission spectrum, and is once again characteristic of the atomic vapour used. A further feature of emission spectra is that they frequently contain regions of continuous emission, where the intensity of the emission varies smoothly with wavelength.

There are important similarities between the emission and absorption spectra of atoms. Emission spectra generally contain far more lines than the corresponding absorption spectra, but there are many lines which are of common wavelength in both, and in principle all the lines in the absorption spectrum also appear in emission, although possibly with rather low intensity.

So far we have seen that on excitation each element gives rise to a characteristic line spectrum; we may use this as an analytical test for the presence of an element, and this test is important because it can detect very small quantities, and is suitable for analysing trace impurities. Before we can use the spectra to provide information on electronic structure, we must understand how the spectra arise, and this will require a more detailed understanding of the nature of light itself. The idea that light involves wave-motion is familiar to us; we know that light can be diffracted, and interference patterns obtained. Similarly we know that different colours correspond to different wavelengths, and hence to different frequencies, wavelength and frequency being related by the equation

$$\text{wavelength} \times \text{frequency} = \text{velocity}$$

$$\lambda \quad \times \quad \nu \quad = \quad c$$

Wavelength and frequency are clearly wave-like properties of light. But we must now look at some experiments which puzzled theoreticians at the turn of the century, and which seem very hard to reconcile with our views of light as a wave-motion.

C. The Photoelectric Effect

Figure 1.4 shows a simple apparatus for observing the photoelectric effect. An evacuated chamber contains a sample of metal, M, and a plate, P. The plate is maintained at a slight positive potential relative to the metal. The metal is not heated, and so ordinarily no electrons pass from M to P. However, if ultraviolet light is focussed onto the metal, it is found that electrons pass from M to P, and are detected by the ammeter A. The important point is that electrons are only emitted if the frequency of the u.v. light exceeds a certain threshold value, which is a specific property of the metal. If the frequency falls below the threshold value, the current drops to zero, no matter what the intensity of the incident light is.

4

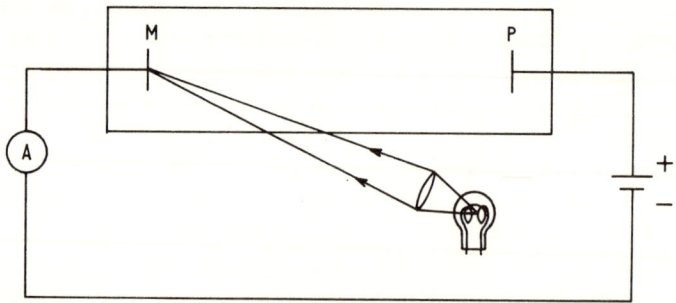

Figure 1.4 Experimental arrangement for observing the photoelectric effect

The photoelectric effect is very hard to explain if we view light as a wave motion. But a return to Newton's corpuscular view of light provides a simple explanation. Light is regarded as comprising a stream of particles, which are called photons. Each photon has an energy E, which is related to the frequency of the radiation ν by the equation

$$E = h\nu \tag{1.1}$$

where h is a constant, called Planck's constant. Now the amount of energy required to remove one electron from the surface of the metal is called the 'work function', and the magnitude of the work function varies from metal to metal. Einstein suggested that electrons could be emitted from the metal only when the energy of one photon, $h\nu$, exceeded the work function for the metal. This explanation implies that a single quantum of energy must be transferred to a single electron, and that the energy of the light is localized and not spread out across a wave front. It is then clear that if the frequency of the light is too low, then the energy of the photons is less than the work function for the metal, and the electrons do not acquire sufficient energy to escape from the surface. The important point is that each electron is emitted by the action of one incident photon; it is impossible under ordinary circumstances for an electron to absorb two photons, each corresponding to a frequency just below the critical frequency.

D. Wavelength, Frequency and Energy

The fundamental equation of Planck

$$E = h\nu$$

was produced as a result of a consideration of black-body radiation. It is important to have a qualitative feeling for what it means. Essentially the energy of a photon is directly related to its frequency and inversely related to its wavelength. Thus blue light is more energetic than red light and ultraviolet light more powerful than visible light. Hence ultraviolet light causes sunburn and fading of dyestuffs and even·shorter wavelengths become increasingly damaging in the form of X-rays or gamma rays. Infrared light is less energetic than visible light and is detected as heat.

In general the wavelength distribution of light emitted by a hot body is such that the hotter the body the more energetic is the bulk of the radiation emitted. Thus an electric fire emits mostly heat with some red light whereas a much hotter lamp filament emits white light covering all of the visible spectrum.

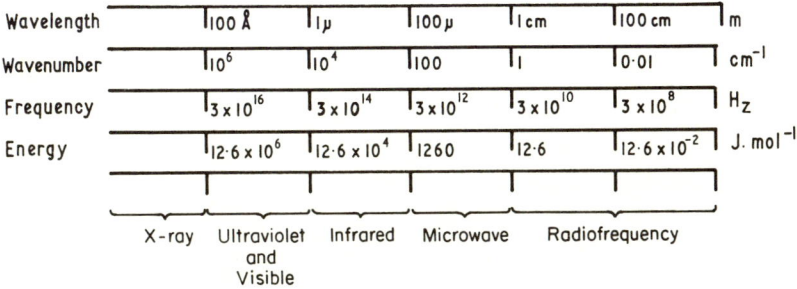

Figure 1.5   The electromagnetic spectrum

Figure 1.5 shows the electromagnetic spectrum.

Much of the contents of this book will be concerned with energy level diagrams. These have energy on a vertical scale and wide gaps between levels will correspond to large energy differences, high frequencies and short wavelengths.

E. Wave-Particle Duality

Further evidence in favour of Einstein's interpretation of the photo-electric effect came from an experiment performed by Compton, who was concerned with the problem of whether photons possessed linear momentum. Compton was able to scatter electrons with a beam of X-rays, and he showed that the laws of conservation of energy and momentum held during this process, as they should do for collisions between particles. Furthermore, he was able to show that in a scattering experiment, momentum was transferred in discrete amounts, rather than continuously.

These experiments, and several others, left the uncomfortable situation where some experiments were to be explained by thinking of radiation as being wave-like, and others by considering radiation as a stream of particles. The balance was redressed somewhat when it was shown that electrons, which had always been considered as particles, could be diffracted by the layers of atoms in regular crystals; it is now accepted that the behaviour of both electromagnetic radiation and electrons is sometimes best described by a wave model, and on other occasions by a particle model.

We may now return to the absorption and emission of light by atoms, where we shall think of light as a stream of photons. An atom can only absorb certain particular wavelengths of light, and these wavelengths each correspond to a particular photon energy. It is therefore clear that an atom can only exist in a discrete number of possible states, and each of these has a well-defined energy. An atom is normally found in the state of lowest energy, and it can 'jump' to another state by absorbing a single photon. Note that, just as in the photoelectric effect, it is not normally possible for an atom to absorb two photons, each with half the required energy. The absorption spectrum therefore contains jumps, or transitions, from the state of lowest energy, the ground state, to various other allowed 'excited' states.

In the emission experiment, the atoms are excited by an electric discharge, or some other means, and are initially in various excited states. In an electric discharge, atoms are excited by collisions with charged particles accelerated in an applied electric field. They can lose energy by emitting photons, and their final state may be the ground state. In this case the photon energy is equal to the energy difference between the excited and ground states, and so the line in the emission spectrum corresponds to a line in the absorption spectrum. Alternatively, the final state may not be the ground state, but a different excited state; the photon energy will not now correspond to a line in the absorption spectrum. Figure 1.6 shows the allowed states, and the emission and absorption spectra, for a simple case.

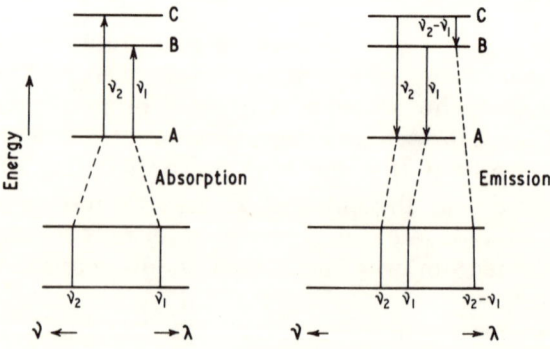

Figure 1.6   Absorption and emission processes

Analysis of experimental spectra allows us to construct energy level diagrams such as that shown in Figure 1.6 for a wide variety of atoms. Normally it is not possible, however, to spot any simple relationships between the various allowed energy levels. The exception to this generalization is the hydrogen atom; in this chapter and the next we shall look at the spectrum of the hydrogen atom in some detail, and this will provide us with enough new information to return to the spectra of more complex atoms.

E. Bohr and the Hydrogen Atom

When an electric discharge is run in a sample of hydrogen gas, the light emitted by the discharge is found to consist of a series of discrete wavelengths, and is the emission spectrum of the hydrogen atoms which are formed. This spectrum consists of several series of lines; one series lies in the visible region, and is called the Balmer series, after its discoverer. Other series lie in the ultraviolet region (Lyman series) and the infrared (the Paschen, Brackett and Pfund series). Long before there was any explanation of how the series were formed, Balmer was able to show that the frequencies of the lines in his series could be fitted to a general formula

$$E = h\nu = R \left(\frac{1}{4} - \frac{1}{n^2} \right) \tag{1.2}$$

where ν is the frequency, R is a constant, and n is an integer, which could take any value from 3 to infinity. Furthermore it was shown that the lines in the Lyman series could be fitted to the equation

$$E = h\nu = R \left(\frac{1}{1} - \frac{1}{n^2} \right) \tag{1.3}$$

where R has the same value as in equation (1.2), and where n can take any integral value from 2 to infinity. Further work on the lines in the infrared showed that all the lines in the hydrogen emission spectrum could be fitted to the general equation

$$E = h\nu = R \left(\frac{1}{n_1^2} - \frac{1}{n_2^2} \right) \tag{1.4}$$

where R is called the Rydberg constant, and n_1 and n_2 are both integers. For a given series n_1 is constant; thus in the Lyman series $n_1 = 1$, and in the Paschen series $n_1 = 3$. Figure 1.7 shows a schematic representation of the emission spectrum of the hydrogen atom. In a given series, as n_2 increases, the separation between adjacent lines

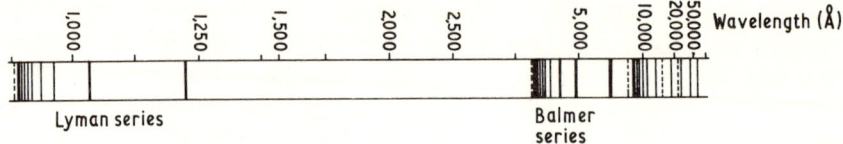

Figure 1.7   The emission spectrum of hydrogen

decreases, and the lines converge on a limiting frequency which has the value

$$E_{\text{limit}} = h\nu_{\text{limit}} = R/n_1^2$$

This frequency is called the series limit, and in principle an infinite number of lines lie at the series limit.

We concluded in the last section that each line in the atomic spectrum corresponded to a transition between two allowed energy levels; from equation (1.4) we may deduce that each of the allowed levels in the hydrogen atom may be described by an integer, n, and that the energy of that level is given by

$$E = -\frac{R}{n^2}$$

Thus if an atom undergoes a transition from a state n_1 to a state n_2, the energy emitted will be the difference in energies between the states

$$E = -\frac{R}{n_2^2} + \frac{R}{n_1^2}$$

But the energy E of the photon emitted is related to its wavelength by the formula

$$E = h\nu \qquad \text{(from equation 1.1)}$$

$$\therefore h\nu = -\frac{R}{n_2^2} + \frac{R}{n_1^2}$$

$$\therefore \nu = \frac{R}{h}\left(\frac{1}{n_1^2} - \frac{1}{n_2^2}\right)$$

which is the experimental observation summarized in equation (1.4).

Bohr was able to explain the form of equation (1.4) in terms of a simple model which rested on three initial postulates. They were

(1) The electron in a hydrogen atom moves in a circular orbit around the positively charged nucleus.

(2) Only certain discrete orbits are allowed, and the electrons do not emit radiation when they are in these orbits.

(3) A single photon is emitted or absorbed when an electron jumps from one allowed orbit to another.

These postulates seemed somewhat arbitrary when they were first proposed; in particular the second postulate is contrary to classical theory. An orbiting electron is continuously accelerating towards the nucleus and according to classical electrodynamics should radiate and lose energy. Bohr's great success was that he made his model quantitative, and obtained astonishing agreement with experiment. Figure 1.8 shows the circular orbit of an electron in a hydrogen atom.

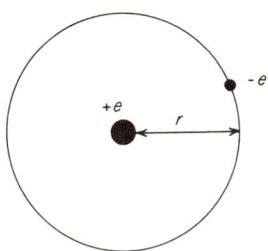

Figure 1.8   Bohr's model for the hydrogen atom

If the electron has mass m and velocity v, then the force required to maintain it in circular orbit is, from classical mechanics,

$$\frac{mv^2}{r}$$

This force arises from the electrostatic attraction of the negative electron and the positive proton, and so

$$\frac{e^2}{4\pi\epsilon_0 r^2} = \frac{mv^2}{r} \tag{1.5}$$

Bohr made the revolutionary assumption that the angular momentum of the electron (mvr) could only have certain values, and that it was quantized in units of $h/2\pi$, where h is again Planck's constant.
    Thus

$$mvr = \frac{nh}{2\pi} \qquad \text{where} \quad n = 1, 2, 3, \ldots \tag{1.6}$$

Equations (1.5) and (1.6) may be combined to give a value of r, the radius of the orbit, in terms of known fundamental constants

$$r = \frac{\epsilon_0 n^2 h^2}{\pi m e^2} \tag{1.7}$$

This gives a value for r for the ground state ($n = 1$) of 0.52917Å ($1\text{Å} = 10^{-10}\text{m}$), which is in itself something of a triumph; viscosity measurements of hydrogen gas suggest that the H_2 molecule has a diameter of about 1Å. This value of r is called the Bohr radius, a_0.

However, even more striking is the expression for frequencies of the lines in the H atom spectrum. The energy of the H atom is given by

E = kinetic energy + potential energy

Kinetic energy = ½ mv^2

Potential energy = work needed to remove electron to infinity

$$= \int_r^\infty (-e^2/4\pi\epsilon_0 r^2)dr = [e^2/4\pi\epsilon_0 r]_r^\infty$$

$$= -e^2/4\pi\epsilon_0 r$$

$$\therefore E = \frac{mv^2}{2} - \frac{e^2}{4\pi\epsilon_0 r} \qquad (1.8)$$

Combining equations (1.5) and (1.8) we obtain

$$E = \frac{e^2}{8\pi\epsilon_0 r} - \frac{e^2}{4\pi\epsilon_0 r}$$

$$= -\frac{e^2}{8\pi\epsilon_0 r}$$

$$= -\left(\frac{me^4}{8\epsilon_0^2 h^2}\right) \cdot \frac{1}{n^2}$$

$$= -(\text{constant}) \times \frac{1}{n^2}$$

Thus the energy emitted in a transition from state n_1 to state n_2 is

$$\Delta E = E_2 - E_1 = h\nu = \text{constant} \times \left(\frac{1}{n_1^2} - \frac{1}{n_2^2}\right) \qquad (1.9)$$

The value of the constant in equation (1.9) may be calculated from the known values of m, e and h, and is found to equal the experimentally determined Rydberg constant almost exactly — a remarkable result.

G. Extensions of the Bohr Theory

Just as the moon does not rotate around the earth, but rather both rotate about their common centre of gravity, so too the electron rotates about the common centre of gravity of the electron and the proton. We can include this factor in our calculations by replacing the electronic mass by the reduced mass of the system μ. μ is given by

$$\mu = \frac{M \times m}{M + m}$$

As the electron is roughly 1800 times less massive than the proton (mass M) the correction is relatively small, but nevertheless brings the agreement between the calculated and experimental Rydberg constants to better than 1 part in 10^4. Furthermore, the correction accounts for small differences between the spectra of hydrogen and the heavier isotope deuterium.

The Bohr theory can also be extended to other one-electron systems such as the He^+ and Li^{2+}. The electrostatic attraction between the nucleus and electron is now Ze^2/r^2, where Ze is the appropriate nuclear charge, and the energy of the states may be shown to be

$$E = -\left(\frac{me^4}{8\epsilon_0^2 h^2}\right) \frac{Z^2}{n^2}$$

(see problem 1.4).

As well as spectral line frequencies, it is also possible to calculate ionization potentials for one-electron atoms. The ionization potential is the energy required to remove an electron from the atom to infinity, and corresponds to exciting the electron to an orbit with quantum number ∞.

$$\therefore \text{Ionization Potential} = R \left(\frac{1}{n_1^2} - \frac{1}{n_\infty^2}\right)$$

$$= R/n_1^2$$

In the spectrum, such an ionization will lead to continuous absorption of energy at frequencies above the ionization limit; this is shown in Figure 1.9.

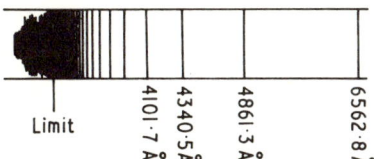

Figure 1.9 The ionization limit in the hydrogen spectrum

H. Failure of the Bohr Theory

In Bohr's simple theory we have seen that there is a single quantum number, n. This quantizes angular momentum,

$mvr = nh/2\pi$

and also energy,

$E = -R/n^2$

Its success with one-electron systems was considerable; but it failed completely with more complicated systems, in particular the alkali metal spectra. Figure 1.10 shows a representation of the absorption spectrum of Na; it shows some similarities to the hydrogen spectrum, but many of the lines appear to be doubled, unlike the hydrogen spectrum, and there is no simple relationship such as the Bohr formula connecting the lines. Not only does the Bohr theory fail to explain the quantitative features of the Na spectrum, it can give no explanation of many of the qualitative features either.

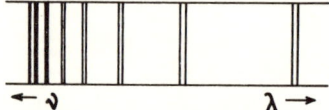

Figure 1.10   Part   of   the spectrum of sodium showing the doubled lines (not to scale)

Sommerfeld extended Bohr's theory by having two quantum numbers. As before, the energy was quantized by n, but now the angular momentum was quantized by a new quantum number k,

$mvr = kh/2\pi$

k took integral values, but was not allowed to exceed n. The explanation of the double lines in the sodium spectrum in theory which embraces the Bohr theory was achieved by assuming that the electrons in atoms have elliptical rather than circular orbits. The Sommerfeld theory gave a good account of the spectra of the alkali metal atoms, though it proved almost impossible to extend to other atoms, and did give some wrong answers. We need not concern ourselves further with this theory, for it was soon discarded, and along with it the so-called 'Old Quantum Theory'.

It seems probable that Sommerfeld's theory could have been improved by some further tinkering with its postulates, but the reason for its abandonment was much more fundamental and philosphical. The whole edifice of the Bohr theory was too arbitrary, and its results too definite. The problem arose from the work of Heisenberg, who produced the famous Uncertainty Principle. His detailed considerations of the workings of a hypothetical light microscope being used to observe microscopic phenomena, such as the motion of an electron, showed that it was not possible *in principle* to know simultaneously the position and momentum of an electron with complete accuracy; the more accurately the one was determined, the less information could be obtained about the other, even with a theoretically perfect instrument. This was of course in direct contradiction to the Bohr and Sommerfeld theories, in which the electrons were viewed as having well-defined orbits, in which the positions and momenta of the electrons were always well known.

The next chapter is concerned with wave mechanics, which came to replace the old quantum theory. It takes account of the work of Heisenberg, and is able to give a satisfactory account of the spectra of all atoms, and not merely some of the more simple cases. Wave mechanics is fundamental to our comprehension of the electronic structure of atoms; nervertheless it is possible to understand much of atomic spectra without a detailed knowledge of wave mechanics, and some readers may prefer to omit Chapter 2, with its necessarily mathematical emphasis, on a first reading.

II

Wave Mechanics

In wave mechanics, which replaced the old quantum theory, the electron is ascribed wave-like properties, and therefore must be described by a wave equation, just like the waves of light or sound. This equation was given by Schrödinger in 1926, and may be taken as the starting point for a discussion of modern quantum theory. It is the generalization of a simple wave equation postulated by de Broglie in 1924. In the old quantum theory, the idea of quantized energy levels and quantized angular momentum is somewhat artificially grafted onto classical physics; in wave mechanics, this quantization is the consequence of some reasonable assumptions about the possible solutions of the Schrödinger equation.

A. The Schrödinger Wave Equation

The Schrödinger wave equation is the starting point for wave mechanics; it may not be deduced from classical mechanics, nor may it be proved in any way, other than retrospectively as its predictions are confirmed by experiment. It is commonly written as

$$\frac{1}{2}\left(\frac{\partial^2}{\partial x^2} + \frac{\partial^2}{\partial y^2} + \frac{\partial^2}{\partial z^2}\right)\Psi + (E - V)\Psi = 0 \tag{2.1}$$

It is possible, however, to give some rationale to the form of this important equation, which was not dreamed up in a flash of insight, but rather follows from some of the work discussed in the last chapter.

As we have seen, Planck gave the energy quantization condition formula

$$E = h\nu$$

which relates the energy E to a wave-like property, the frequency ν. Einstein, from his work on relativity, produced the even more celebrated equation

$$E = mc^2$$

which relates the energy E to the particle-like properties, the velocity c of a mass m.

De Broglie, in 1924, made the connection between the wave and the particle behaviour, and obtained

$$h\nu = mc^2$$

As momentum $p = mc$, and $\nu = c/\lambda$, this may be written as

$$p = h/\lambda,$$

where the left-hand side is a particle property, momentum, and the right-hand side contains a wave property, wavelength.

Schrödinger's massive contribution was to recast the de Broglie relation into a form which is more useful, although it appears to be more complicated, and to postulate its applicability to situations where particles have not only kinetic but also potential energy, V.

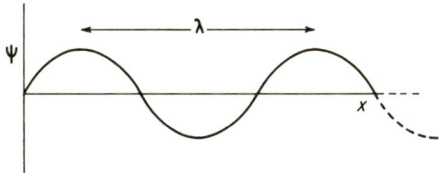

Figure 2.1   A simple standing wave

To obtain the common form of the Schrödinger equation, we may start with the equation for the amplitude of a simple standing wave (Figure 2.1),

$$\Psi = A \sin 2\pi x/\lambda$$

where Ψ is just the quantity whose value varies with distance in a wave-like manner. Differentiating once, and then again, we obtain:

$$\frac{d\Psi}{dx} = \frac{2\pi}{\lambda} A \cos 2\pi x/\lambda$$

and

$$\frac{d^2\Psi}{dx^2} = -\frac{4\pi^2}{\lambda^2} A \sin 2\pi x/\lambda = -\frac{4\pi^2}{\lambda^2} \Psi$$

If we denote the kinetic energy as T, then for a particle of mass m and velocity c,

$$T = \frac{1}{2} mc^2 = \frac{1}{2m} p^2$$

Using the de Broglie relation,

$$T = \frac{1}{2m} \frac{h^2}{\lambda^2}$$

and therefore

$$\frac{d^2 \Psi}{dx^2} = -\frac{8\pi^2 m}{h^2} T \cdot \Psi \tag{2.2}$$

This equation applies only to a particle moving in field-free space, where the particle has only kinetic energy. If the particle has potential energy V then the total energy is given by $E = T + V$. Schrödinger postulated that for a system that does not change with time, the kinetic energy T in equation (2.2) could be replaced by $(E - V)$. Thus the Schrödinger equation for one dimension may be written as

$$\frac{d^2 \Psi}{dx^2} + \frac{8\pi^2 m}{h^2} (E - V)\Psi = 0$$

In three dimensions this equation becomes

$$\frac{\partial^2 \Psi}{\partial x^2} + \frac{\partial^2 \Psi}{\partial y^2} + \frac{\partial^2 \Psi}{\partial z^2} + \frac{8\pi^2 m}{h^2} (E - V)\Psi = 0$$

This equation looks cumbersome, but this is partly due to the units employed; it becomes simpler if we use so-called atomic units. The electronic mass m and charge e are put equal to unity, and the unit of length is chosen to be the Bohr radius a_0 (see Chapter 1); in these units $h = 2\pi$. If we further denote double differentiation with respect to each of the three coordinates by ∇^2, the equation becomes

$$\tfrac{1}{2} \nabla^2 \Psi + (E - V)\Psi = 0$$

which is the form given in equation (2.1).

B. Meaning and Properties of Ψ

The quantity Ψ which we have described as varying in a wave-like manner has not been very clearly defined; it represents the amplitude of the electron wave, and is called the 'wave function'. Its interpretation is due mainly to Born, who by analogy with other wave equations considered Ψ^2 as a probability. For a system containing one particle, $\Psi^2 dx$ (or more correctly, $\Psi^* \Psi dx$ if Ψ is a complex function, and Ψ^* is its complex conjugate) is the probability of finding the particle in the region defined by dx. Alternatively we can envisage $\Psi^* \Psi dx$ as being a measure of the density of matter in the region defined by dx, with the particle having no discrete nature; it is not strictly correct to consider 'particle clouds', with the particles having no discrete nature, but nevertheless it provides a convenient and generally accurate way of thinking of wave functions. This interpretation of Ψ^2 as a probability enables the new quantum mechanics to incorporate Heisenberg's

Uncertainty Principle in a manner which was impossible with the old theory.

Before we are in a position to solve the Schrödinger wave equation for any simple cases, we must first impose some very reasonable conditions on Ψ that follow as a direct result of the probability interpretation.

1. Ψ must always be finite at any point

If Ψ were to become infinite at any point in space, this would correspond to a certainty of finding the particle there, which would conflict with the uncertainty principle.

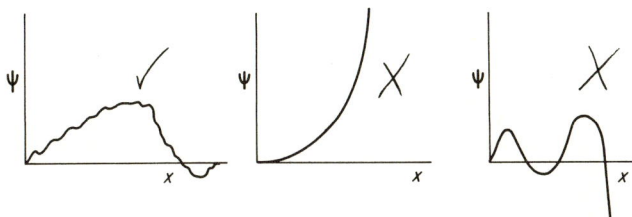

Figure 2.2   Allowed and forbidden wave functions: Ψ must be finite

2. Ψ must be single-valued at any point in space

There can clearly be only one value of the probability of finding the particle at a given point in space.

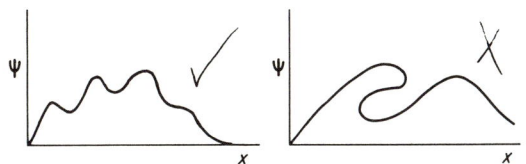

Figure 2.3   Allowed   and   forbidden   wave functions: Ψ must be single valued

3. Ψ must be continuous

If the probability of finding a particle at point x has a given value, that at $(x + \delta x)$ must be similar as $\delta x \to 0$. If this were not so, the double differentiation in the Schrödinger equation could not be performed. An

18

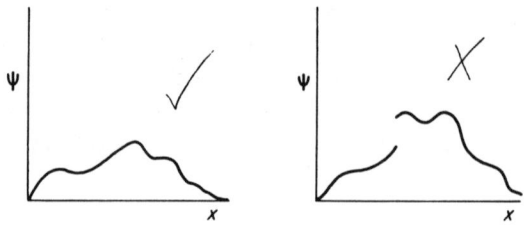

Figure 2.4   Allowed and forbidden wave functions: Ψ must be continuous

important consequence of this condition is that Ψ must become zero at infinity.

C. Solution of the Schrödinger Wave Equation

If we wish to obtain the wave function for a particle in a particular defined environment, the procedure is now as follows:

(i) The differential Schrödinger equation is set up in a suitable set of coordinates (cartesian x,y,z for problems where this is simplest, polar r,θ,ϕ for problems with spherical symmetry, see Figure 2.5)

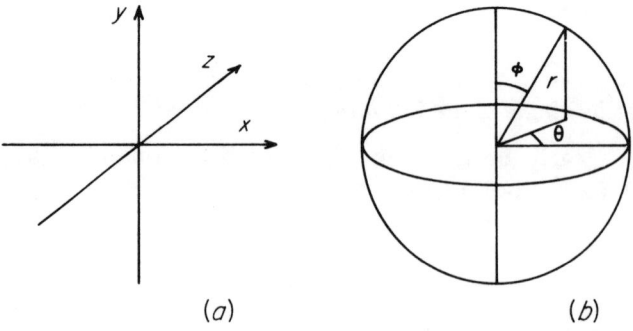

(a) $\qquad\qquad\qquad\qquad\qquad\qquad$ (b)

Figure 2.5 (a) Cartesian (x, y, z) coordinates (b) Polar (r, θ, ϕ) coordinates

(ii) The potential V is introduced in functional form

(iii) The differential equation is solved subject to the conditions given in the previous section

(iv) From this a set of wave functions ψ_1, ψ_2, . . . are obtained, with energies E_1, E_2, ... These are quite simple mathematical functions, such as sine curves, which satisfy the Schrödinger wave equation.

The surprising and exciting consequence of this calculation is that the reasonable limitations that we have set on the wave functions lead naturally to the quantization of energy and angular momentum. This may be seen by considering the solution of the Schrödinger wave equation for a few simple cases.

D. The Particle in a One-dimensional Box

The simplest problem that we can consider is that of a particle in a one-dimensional box where the potential energy V is zero within the box, but infinite outside the box (Figure 2.6).

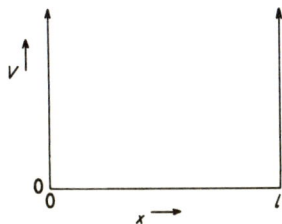

Figure 2.6   Potential energy (V) of a particle in a one-dimensional box

The Schrödinger equation which has to be solved to obtain the allowed energy levels of this system, and the corresponding allowed wave functions, or eigenfunctions as they are called, is (in atomic units)

$$\frac{d^2 \Psi}{dx^2} + 2(E - V)\Psi = 0 \tag{2.3}$$

By inspection of equation (2.3), the value of Ψ must be zero if $V = \infty$ for finite values of the energy E. The only acceptable solutions therefore have $\Psi = 0$ at $x = 0$ and $x = l$, and vary smoothly between these limits. Functions which satisfy these conditions can be expressed in the form:

$$\Psi_n = A \sin \frac{n\pi x}{l}$$

as can be verified by substitution in equation (2.3).

The subscript n designates the particular solution, and n is an integral number; the presence of the boundary conditions gives rise to a quantum number. These wave functions are shown in Figure 2.7. If the wave functions are substituted back in the Schrödinger equation, then an expression for their energies is obtained:

$$E_n = \frac{n^2 h^2}{8ml^2}$$

This gives rise to an energy level diagram of the type shown in Figure 2.8.

20

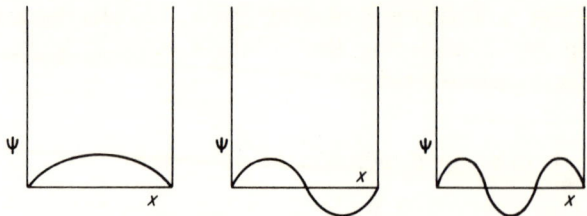

Figure 2.7 Allowed wave functions (eigenfunctions) for the particle in a one-dimensional box

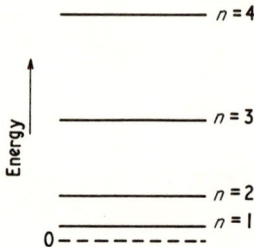

Figure 2.8 Energy levels of particle in a one-dimensional box

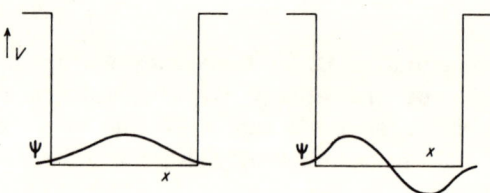

Figure 2.9 Wave functions for a particle in a one-dimensional box with finite external potential, V.

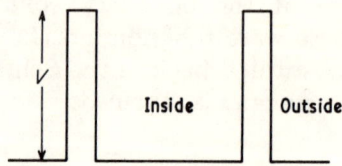

Figure 2.10 Potential for a 'box' with finite barrier

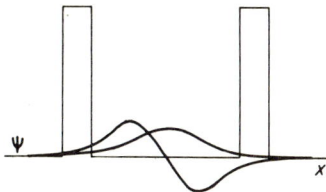

Figure 2.11 Wave functions for a particle in a finite-sided box

This problem can be made more realistic by allowing the potential energy outside the box to have a large, rather than an infinite value. The wave function now goes to zero not at the edge of the box, but at infinity, and the allowed solutions of the Schrödinger equation are illustrated in Figure 2.9. It may be seen that there is some finite chance of finding the particle outside the box, since Ψ does have a finite, if small, value just beyond the boundary. If the box has its boundary in the form of a potential barrier, as shown in Figure 2.10, then solutions of the type shown in Figure 2.11 are possible. These clearly imply that there is a finite possibility of finding the particle outside the box, even though it does not have sufficient energy to surmount the potential barrier which holds it in the box. This is the important *tunnel effect* which is observed experimentally. One of the most striking examples is the emission of particles from the nuclei of radioactive atoms such as uranium; the particles tunnel through the potential barrier which holds the constituent parts of the nucleus together.

E. Applications of the Particle in a Box Solution

Although the problem of a particle in a one-dimensional box is artificial, and only a mathematical model, it can serve as the basis for a number of semi-quantitative explanations of observed phenomena. It may seem far removed from organic chemistry, but long-chain conjugated hydrocarbons have very mobile electrons which are in a situation similar in some ways to the particle in a box case (Figure 2.12). The formula for the energy levels of the electrons in the

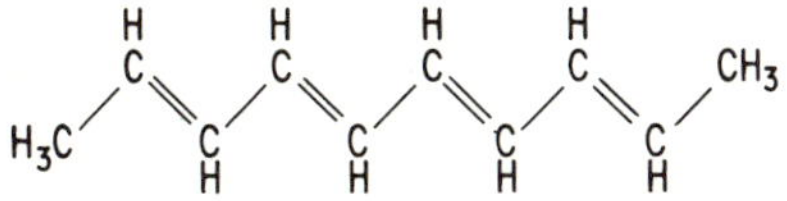

Figure 2.12 A long-chain conjugated hydrocarbon

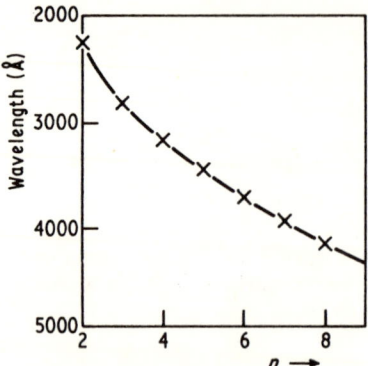

Figure 2.13 Wavelength of the first absorption band of conjugated hydrocarbons

hydrocarbon has been obtained in the last section:

$$E_n = \frac{n^2 h^2}{8ml^2}$$

it implies that the energy separation between the lowest two levels should decrease as l increases; (see problem 2.4).

Thus the absorption of light should be increasingly towards the red end of the spectrum as the molecules get larger; this is indeed the case, and is illustrated in Figure 2.13.

F. The Particle in a Two-Dimensional Box

So far we have considered the situation where a particle is held in a one-dimensional box. Extending this analysis to two and three dimensions adds no great extra complexity, while adding markedly to the realism of the problem.

The Schrödinger equation for a particle in a rectangular two-dimensional box with infinite walls is:

$$\frac{\partial^2 \Psi}{\partial x^2} + \frac{\partial^2 \Psi}{\partial y^2} + 2E\Psi = 0$$

This equation may be separated into two one-dimensional problems, each of which has the familiar solution

$$E_{x,n_x} = \frac{n_x^2 h^2}{8ml_x^2}$$

$$E_{y,n_y} = \frac{n_y^2 h^2}{8ml_y^2}$$

and the total energy is given by

$$E_{n_x, n_y} = \frac{h^2}{8m} \left(\frac{n_x^2}{l_x^2} + \frac{n_y^2}{l_y^2} \right)$$

Because this is a two-dimensional problem, there are two quantum numbers; the extension to three dimensions is clear, and a particle in a three-dimensional box would be expected to have three quantum numbers.

The wave function of the particle in a two-dimensional box is a product of two terms, one involving x and the other y.

$$\Psi_{n_x, n_y} = A \sin \frac{n_x \pi x}{l_x} \cdot \sin \frac{n_y \pi y}{l_y}$$

If the box is square, that is if $l_x = l_y$, the wave function reduces to

$$\Psi_{n_x n_y} = A \sin \frac{n_x \pi x}{l} \cdot \sin \frac{n_y \pi y}{l}$$

and

$$E_{n_x n_y} = \frac{h^2}{8ml^2} (n_x^2 + n_y^2)$$

These wave functions are shown in Figure 2.14.

The first energy level has $n_x = n_y = 1$ (remember that n cannot equal zero, as the wave function would then always be zero); however the second energy level can have $n_x = 2$, $n_y = 1$, or $n_x = 1$, $n_y = 2$. This energy level is said to be doubly degenerate. The energy levels of the

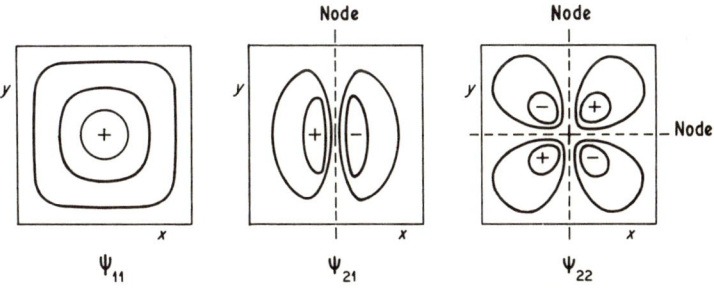

Figure 2.14   Wave functions for a particle in a two-dimensional box represented as contour diagrams

24

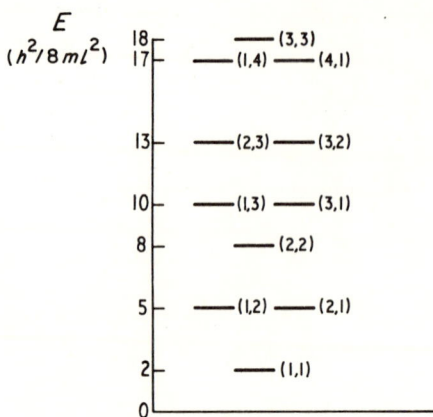

Figure 2.15 Energy levels of particle in a two-dimensional box

particle in the two-dimensional box are shown in the energy level diagram in Figure 2.15.

G. The Hydrogen Atom

The hydrogen atom is clearly most suitably approached using spherical polar coordinates r, θ and ϕ since it is a spherically symmetrical system. The Schrödinger equation is:

$$\tfrac{1}{2}\nabla^2 \Psi + (E - V)\Psi = 0$$

The potential V is the electrostatic nuclear-electron attraction $-e^2/4\pi\epsilon_0 r$ or, in atomic units, $-1/r$. Thus

$$\{-\tfrac{1}{2}\nabla^2 - 1/r\}\Psi = E\Psi$$

Just as the particle in a two-dimensional box problem could be separated into three equations, in x, y and z, this differential equation can be separated if we suppose that the wave function $\Psi(r, \theta, \phi)$ is a product:

$$\Psi(r, \theta, \phi) = R(r)\,\Theta(\theta)\,\Phi(\phi)$$

We then have three equations to solve subject to the limitations of acceptability of the behaviour of the wave functions, and we shall expect to encounter three quantum numbers as our problem involves three coordinates.

The actual mathematical steps involved in the solution of the differential equations are straightforward but lengthy, and we will not rehearse them here. The full derivation is given in many books (see the list of suggestions for further reading). It turns out however that the mathematical functions $R(r)$, $\Theta(\theta)$ and $\Phi(\phi)$ are already well-known to mathematicians, and that each coordinate does give rise to a quantum number; these quantum numbers are designated n, l and m_l for the r, θ and ϕ parts of the eigenfunctions. The actual shapes of the allowed wave functions we shall discuss in the next chapter.

III

The Hydrogen Atom and the
Periodic Table

In the last chapter we introduced the Schrödinger wave equation, whose solution allows us to obtain a wave description of a particle in a defined environment. The particle is described by a 'wave function', Ψ, which is simply the amplitude of the wave; the quantity $\Psi^2 dx$ represents the probability of finding the particle in the space dx. Knowledge of the wave function of a particle allows us to build up a picture of a 'particle cloud'. We also showed that wave functions can be obtained for the electron in a hydrogen atom, and that the hydrogen atom wave functions require three quantum numbers, designated n, l and m, to describe them. These wave functions are often called orbitals.

We must now discuss these three quantum numbers in more detail and consider the shapes of the allowed functions, or orbitals. We shall then see that these provide a basis for describing not only the electronic structure of the hydrogen atom, but also all other atoms, and we shall be able to understand the composition of the entire Periodic Table of elements simply in terms of the allowed hydrogen wave functions.

A. Quantum Numbers in the Hydrogen Atom

The quantum number n is called the principal quantum number, and corresponds exactly to Bohr's n in that it specifies the energy of the atom. As in Bohr's theory,

$$E_n = -R/n^2$$

A mathematical function is said to have a *node* when it changes sign. Hydrogen atom wave functions are three dimensional, and may thus have nodes in the r (radial) direction, or angular nodes which will pass through the origin. (Remember that although Ψ^2, which represents a probability, must be positive, Ψ may be positive or negative.) It is found that the total number of nodes in the hydrogen wave functions is equal to $(n-1)$: this may be verified by inspecting the diagrams in the next section.

The quantum number l is called the azimuthal quantum number; it is equal to the number of angular nodes and has a maximum value of $(n-1)$. All integral values of l from 0 to $(n-1)$ are allowed. This quantum number also specifies the orbital angular momentum of the electron, p_θ.

$$p_\theta = \sqrt{l(l+1)}h/2\pi$$

Thus angular momentum is quantized as well as energy.

For historical reasons, electrons which have values of $l = 0, 1, 2, 3, 4,$ are designated s, p, d, f, g. Thus if an electron is in an orbital with $n = 2$ and $l = 1$, it is called a $2p$ electron; if $n = 2$ and $l = 0$, it is called a $2s$ electron.

The third quantum number which comes from the solution of the Schrödinger equation for the hydrogen atom, m_l, is called the magnetic quantum number. It refers to the orientation of the orbital, and specifies the component of the angular momentum in a particular direction. This direction is normally defined by a magnetic field.

The origin of this quantization of the component of the angular momentum can be understood by a classical analogy. As the electron moves round the stationary nucleus, with an angular momentum defined by the l quantum number, it produces a magnetic moment, as it is a charged particle. The magnet thus produced will precess around a magnetic field, just as a gyroscope precesses round the earth's gravitational field.

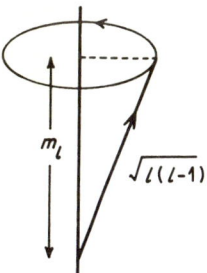

Figure 3.1   The component of an angular momentum vector

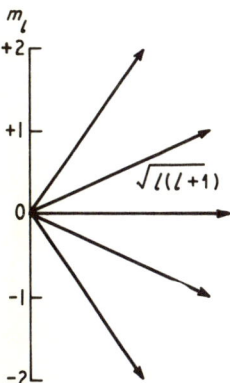

Figure 3.2   Possible m_l values for an orbital angular momentum $l = 2$

The component of the angular momentum in the direction of the magnetic field, the z direction, cannot have *any* value, but is quantized in units of $h/2\pi$.

$$p_{\theta_z} = m_l(h/2\pi)$$

Since the component of a vector cannot have a larger value than the vector itself, m_l can take the integral values $-l$, $(-l+1)$ ... $(l-1)$, l, and thus has $(2l+1)$ possible values. The case where $l = 2$ is illustrated in Figure 3.2; m_l may take five possible values.

B. Wave Functions of the Hydrogen Atom

The actual hydrogen wave functions are difficult to illustrate in two dimensions as they are themselves three-dimensional, but the ground state ($n = 1$, $l = 0$, $m_l = 0$) and the low-lying energy levels have relatively simple shapes which can be easily envisaged.

1. $n = 1$, $l = 0$, $m_l = 0$

$$\Psi_{1,0,0} = Ne^{-Zr/a_0}$$

where N is a constant, Z is the nuclear charge, and a_0 is the Bohr radius. The radial variation of the wave function is shown in Figure 3.3. In this case, there is no variation of the wave function with θ, and consequently there is no variation with ϕ. Thus if we take the nucleus as the origin, the wave function decays away exponentially, no matter which direction we choose. This is drawn in a two-dimensional perspective of the three-dimensional behaviour as shown in Figure 3.4.

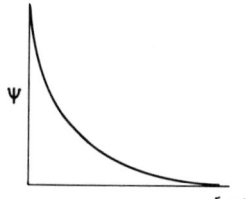

Figure 3.3   Radial variation of a 1s atomic orbital

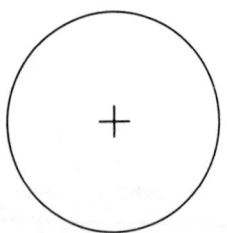

Figure 3.4   Schematic representation of the three-dimensional variation of a 1s atomic wave function

The + sign indicates that the value of Ψ is always positive. The orbital therefore has no nodes, as we should expect as the value of n is 1. It is important to remember that Figure 3.3 represents a plot of Ψ, rather than Ψ^2 against r; in the case of the spherically symmetric 1s function shown here, they both have similar appearance, though this is not generally the case.

2. $n = 2, l = 0, m_l = 0$

As $l = 0$, there are no angular nodes in the 2s wave function, and indeed there is no variation of the wave function with θ or ϕ. Therefore as $n = 2$, there is one radial node. The radial variation of the wave function is shown in Figure 3.5. Again the wave function decays away in this fashion no matter which direction from the nucleus we choose. This may be represented as shown in Figure 3.6.

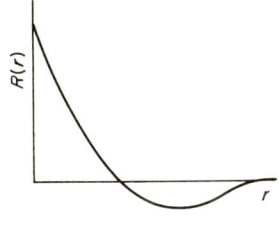

Figure 3.5 Radial variation of an atomic 2s wave function

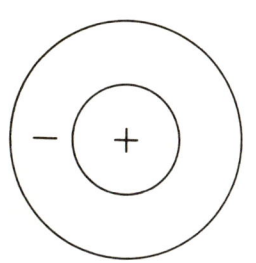

Figure 3.6 Representation of the three-dimensional variation of an atomic 2s orbital

3. $n = 2, l = 1, m_l = 0$

In this case, a 2p function, there is one angular node, as $l = 1$, and therefore no radial node, as the total number of nodes is only one. The wave function is the product of a radial function $R(r)$ and of two angular functions $\Theta(\theta)$ and $\Phi(\phi)$: the variation of $R(r)$ with r may be represented as in Figure 3.7. Note that unlike s orbitals, the value of Ψ drops to zero at the nucleus; only s orbitals have a non-zero electron density at the nucleus, and this has important consequences in magnetic resonance spectroscopy.

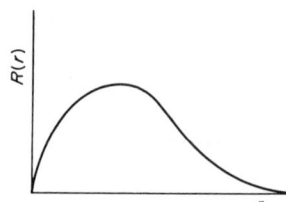

Figure 3.7 Radial variation of an atomic $2p$ wave function

The angular node arises from the dependence of Ψ on θ:

$$\Theta(\theta) = \cos\theta$$

In the θ plane, we have the representation of Figure 3.8. When the function is multiplied by $R(r)$, we obtain a different variation of Ψ with r depending on the precise direction in which we move from the nucleus. This is summarized in one of the well-known drawings of a $2p$ orbital, shown in Figure 3.9. Again there is no dependence of the wave function on ϕ, that is $\Phi(\phi) = 1$.

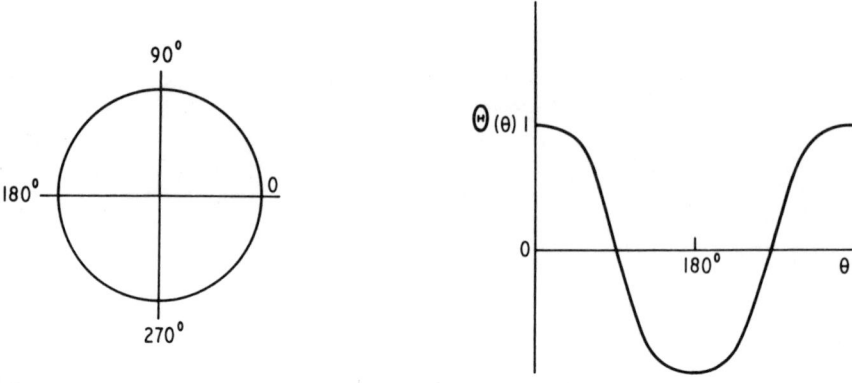

Figure 3.8 Angular dependence of a $2p$ wave function

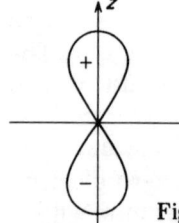

Figure 3.9 Representation of the three-dimensional variation of an atomic $2p$ wave function

4. $n = 2$, $l = 1$, $m_l = \pm 1$

In these two cases, the values of n and l are the same as in the previous example; only the value of m_l has changed. This corresponds to the orbitals having the same shape and size as that in Figure 3.9; only the orientation of the orbital changes. This is often represented by two further $2p$ orbitals, with their lobes directed along the x and y axes* (Figure 3.10)

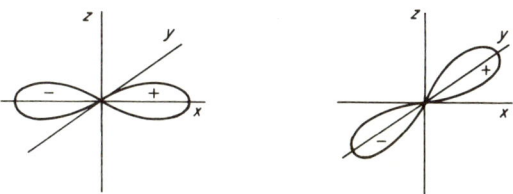

Figure 3.10 $2p_x$ and $2p_y$ orbitals

5. $n = 3$

When $n = 3$, l may take the values 0, 1 or 2. When $n = 0$ or 1 we have $3s$ and $3p$ functions; these have the same angular parts as do $2s$ and $2p$ functions. Their radial parts differ from those for $2s$ and $2p$ functions; they each have one extra radial node. These are shown in Figure 3.11.

If $l = 2$, we have a $3d$ orbital; $3d$ orbitals have a simple radial function, with no nodes, but a more complex angular part. Figure 3.12 shows the radial distribution of a $3d$ orbital.

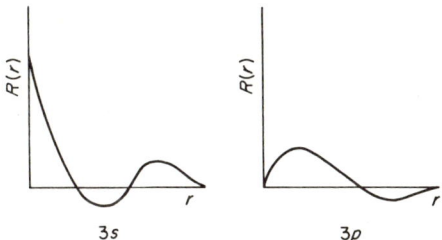

Figure 3.11   Atomic $3s$ and $3p$ radial functions

*This is not strictly correct, as the functions p_x, p_y and p_z that we have shown do not have corresponding angular momenta projections of ± 1 and 0. It is necessary to take complex linear combinations of the p_x and p_y orbitals to obtain the correct functions; nevertheless, this does not detract from the usefulness of these orbitals as chemical models.

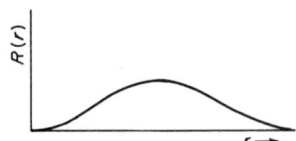

Figure 3.12   Radial dependence of a $3d$ function

C. Energy Levels of the Hydrogen Atom

Solution of the Schrödinger wave equation for the hydrogen atom yields not only the allowed wave-functions Ψ which satisfy the equation

$$\left[\frac{1}{2}\nabla^2 - \frac{1}{r}\right]\Psi = E\Psi$$

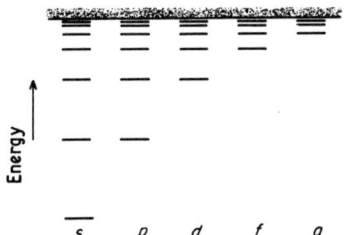

Figure 3.13   Schematic energy level diagram for the hydrogen atom

but also the values of E, which may be labelled by quantum numbers, which are the energies of the various states. The energy level diagram for the hydrogen atom is given in Figure 3.13. It can be seen that all the levels, other than the lowest level, are degenerate, that is consist of two or more states of the same energy. In these diagrams only the vertical scale has any significance, and so it is clear that the Bohr theory with its single quantum number can give a correct prediction of the hydrogen atom spectrum. However, in more complicated cases, such as in polyelectronic atoms or when magnetic fields are applied, these degeneracies are lifted, and then the success of the wave mechanical theory contrasts with the failure of the old quantum theory.

D. Electron Spin — the Fourth Quantum Number

From the examples of solutions of the Schrödinger wave equations that we have seen in Chapter 2, we have become used to the idea that the number of quantum numbers required is equal to the number of dimensions in the problem. Thus in the one-dimensional problem we obtained one quantum number, and in the two-dimensional problem we obtained two quantum numbers. It might therefore be anticipated that three quantum numbers would be sufficient to describe the hydrogen atom. However, the powerful work of Einstein showed the need for four quantum numbers to describe the hydrogen atom fully; in relativity, time is added as a fourth dimension. Now the Schrödinger equation is incomplete in the sense that it takes no account of relativity. Relativistic effects become important when particles move with velocities approaching the speed of light, and this is frequently the case with electrons in atoms. It is therefore necessary to incorporate relativistic theory into the Schrödinger equation, and this was first done by Dirac. Dirac's equations contain four dimensions, and so naturally four quantum numbers appear from this theory.

Historically the need for the fourth quantum number became clear from experience rather than theory, and Uhlenbeck and Goudsmit introduced the electron spin angular momentum s. s has the value $\frac{1}{2}h/2\pi$ [or more correctly $\sqrt{\frac{1}{2}(\frac{1}{2} + 1)}h/2\pi$] and can have components of this intrinsic angular momentum as it spins about its *own* axis, given by the fourth quantum number $m_s = \pm \frac{1}{2}$ corresponding to the two possible directions of spin. This is illustrated in Figure 3.14.

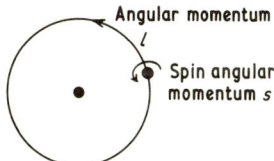

Figure 3.14   Spin and orbital angular momentum

The quantum number m_s was introduced to explain the well-known fact that the spectra of the alkali metals contain closely-spaced pairs of lines, even though such atoms are thought of as consisting of a single electron moving round a core. With the extra quantum number m_s, the total angular momentum can take two possible values $(l + \frac{1}{2})$ and $(l - \frac{1}{2})$.

A very convincing demonstration of the fact that the electron does act as a spinning charged particle, and hence has a magnetic moment, was given by the Stern—Gerlach experiment. We shall discuss this experiment in more detail later, but its essential feature was that a beam of alkali metal atoms, which have no orbital angular momentum, was split into two beams by an inhomogeneous magnetic field, due to the magnetic effect of the two possible values of the spin.

E. Polyelectronic Atoms and the Pauli Principle

It might be thought that it is possible to explain the observed emission and absorption spectra of all atoms simply by writing down the form of the Schrödinger equation appropriate to the problem, separating the equation in radial and angular parts and solving the resultant equations. Unfortunately this is not the case as the problem is not separable owing to the presence of electron-electron repulsion terms. Only in the case of the hydrogen-like atoms, with one nucleus and one electron, is it possible to solve the Schrödinger equation exactly in analytical form.

For other atoms it is necessary to make the so-called 'orbital approximation' if we are to make any headway in understanding electronic structure. This involves writing the wave function of the atom as a product of 'orbitals' or functions which apply individually to each of the electrons. Each electron then has its own set of four quantum numbers n, l, m_l and m_s.

We need now only one more piece of information if we are to be able to rationalize the electronic structures of polyelectronic atoms. This is embodied in the famous Pauli principle. In its more common, though less fundamental, form this principle states that:

'no two electrons in the same atom may have the same set of all four quantum numbers'.

Every pair of electrons in an atom must therefore differ in at least one quantum number.

This form of the Pauli principle is a consequence of a more basic postulate concerned with the symmetry of wave functions. The square of the wave function Ψ^2 is a probability function which tells us about the likelihood of finding electrons at a particular point in space. Clearly, if we label the electrons with indices $1, 2, 3 \ldots n$, and then exchange labels, the electron density cannot change if electrons are all identical particles. Therefore Ψ^2 must be invariant to changes of labelling the electrons, and therefore Ψ must either remain the same, or merely change sign ($\Psi \times \Psi = -\Psi \times -\Psi = \Psi^2$). It is universal in nature that particles with spin ½, such as electrons, behave such that interchange of identical particles causes the wave function to change sign. This is called antisymmetric behaviour.

We can now show how the Pauli principle, stated in its more fundamental form

'Wave functions must be antisymmetric with respect to interchange of electrons'

is equivalent to the form given above. Let us consider the helium atom, which has two electrons. These are both $1s$ electrons in the ground state, that is they both have $n = 1$, $l = 0$, $m_l = 0$. The two values of the m_s quantum number, $\pm\frac{1}{2}$, are usually denoted α and β. The wave function is not then just a simple product

$$\Psi' = 1s^{\alpha}(1)\, 1s^{\beta}(2)$$

where the electrons are labelled (1) and (2), as changing the labels changes the wave function. To obey the Pauli principle, we must take the linear combination

$$\Psi = 1s^{\alpha}(1)\, 1s^{\beta}(2) - 1s^{\alpha}(2)\, 1s^{\beta}(1)$$

If we now exchange the labels (1) and (2), the first term becomes the second, and the second becomes the first; thus Ψ changes into $-\Psi$. Ψ^2 of course remains unchanged. In exactly the same way we may form the wave function for an excited state, where one electron is a $1s$ electron and the other is $2s$.

$$\Psi = 1s^{\alpha}(1)\, 2s^{\alpha}(2) - 1s^{\alpha}(2)\, 2s^{\alpha}(1)$$

As long as there is some part (one quantum number) different between the two electrons, then it is always possible to form a wave function which satisfies the Pauli principle. However, if the two electrons were to have four identical quantum numbers, then the two terms in the wave function would be identical and the wave function would vanish.

In this example we have considered only the helium atom, with its two electrons, but the same procedure could be adopted for any pair of electrons in any atom and the conclusion would be the same; if we are to be able to form an antisymmetric wave function, then no two electrons can have the same four quantum numbers.

F. The Periodic Table

We are now in a position to consider the electronic structures of the ground states of polyatomic atoms. We may begin by starting with the hydrogen atom, in which there is one electron in a $1s$ orbital, and proceed to the helium atom by increasing the nuclear charge by one, and adding one more electron. This electron, as we have seen, also occupies a $1s$ orbital, but has the opposite spin from the first electron, thus satisfying the Pauli principle. The configuration of H is said to be $1s$, and that of He $1s^2$; this method of obtaining the electron configurations is often called the *aufbau* principle.

In He the $1s$ orbital is now full; it can take no more electrons without contravening the Pauli principle. Thus in lithium, which has a total of three electrons, it is necessary to have one electron in an orbital with $n = 2$. In fact the electron occupies a $2s$ orbital, and the configuration of Li is $1s^2 2s$. We can now build up a table of the electron configurations of all the atoms as far as Ne (10 electrons); this is given in Table 3.1.

Table 3.1

H	$1s$	C	$1s^2\,2s^2\,2p^2$
He	$1s^2$	N	$1s^2\,2s^2\,2p^3$
Li	$1s^2\,2s$	O	$1s^2\,2s^2\,2p^4$
Be	$1s^2\,2s^2$	F	$1s^2\,2s^2\,2p^5$
B	$1s^2\,2s^2\,2p$	Ne	$1s^2\,2s^2\,2p^6$

The order in which the sub-levels are filled is given by Hund's rule, which states that the electrons shall have as many parallel spins as the Pauli principle allows. Thus we can assign the quantum numbers of the seven electrons in the nitrogen atom as:

n	l	m_l	m_s
1	0	0	$\tfrac{1}{2}$
1	0	0	$-\tfrac{1}{2}$
2	0	0	$\tfrac{1}{2}$
2	0	0	$-\tfrac{1}{2}$
2	1	-1	$\tfrac{1}{2}$
2	1	0	$\tfrac{1}{2}$
2	1	$+1$	$\tfrac{1}{2}$

This 'rule' is a simple manifestation of the fact that the negatively charged electrons avoid each other as far as possible. However, the most important point is that the electrons occupy the orbitals of lowest energy (that is, most stable orbitals) compatible with the Pauli principle. It is therefore clear that in the elements Li $-$ F the $2s$ orbital is more stable than the $2p$ orbital, whereas in the H atom, the two orbitals had the same energy, that is were degenerate. This is due to the different shielding of the $2s$ and $2p$ orbitals from the nucleus by the inner $1s$ electrons. Because of the forms of the radial distribution of $2s$ and $2p$ orbitals, shown in Figures 3.6 and 3.7, the $2s$ electron is able to penetrate the tightly bound pair of $1s$ electrons better than the $2p$ electron. It sees a charge which is greater than +1 and therefore has a slightly lower energy. There are a number of other examples where the order of the energies of the orbitals does not follow the hydrogenic case; orbitals usually have the following energy sequence:

$$1s < 2s < 2p < 3s < 3p < 4s < 3d < 4p < 5s < 4d < 5p < 6s < 4f$$

and this is illustrated in Figure 3.15.

With the aid of Figure 3.15 we are now able to explain the structure of the Periodic Table; the Periodic Table was originally designed to illustrate similarities and patterns in physical and chemical properties, but these are simply related to electronic configurations, and atoms with similar configurations have similar chemical properties. Thus in the first two short periods, the $1s$, $2s$ and $2p$ orbitals are being filled, as we

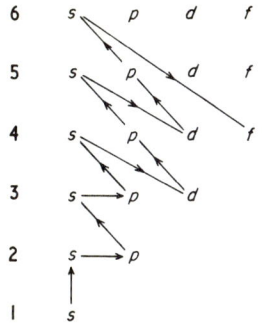

Figure 3.15   Energy sequence of atomic orbitals

can see from Table 3.1. The number of elements in each period follows directly from the possible values of the quantum numbers we obtained from solution of the Schrödinger equation for the hydrogen atom. In the next short period, from Na to Ar, the 3s and 3p orbitals are filled, and the properties of these elements are similar to those of the previous period, where the 2s and 2p orbitals were filled.

After the 3p orbital is filled, we can see from Figure 3.15 that the 4s orbital is now the lowest available orbital, even though the 3d orbital is still unfilled. This is again due to the better penetration of the inner electrons by the 4s orbital than the 3d; their radial distributions are shown in Figure 3.16. The 4s electron is more tightly bound than the 3d electron; it can experience more of the nuclear charge, as it is less well shielded by the inner electrons than the 3d electron. Thus the first two elements of the next period, K and Ca, resemble Na and Mg, having one and two 4s electrons, respectively. But in Sc, the 3d orbital begins to fill, and from Sc to Zn we find the transition metals, which have no counterparts in the earlier periods. After the 3d orbital is full, the 4p orbital fills from Ga to Kr, and these elements are similar to the elements in the previous two periods. The electron configurations of the elements of the 1st long period, from K to Kr, are given in Table 3.2.

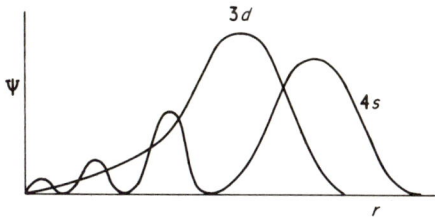

Figure 3.16   Radial distribution of 3d and 4s atomic orbitals

Table 3.2   Electron Configurations of the First Long Period

	1s	2s	2p	3s	3p	3d	4s	4p
K	$1s^2$	$2s^2$	$2p^6$	$3s^2$	$3p^6$		$4s$	
Ca	$1s^2$	$2s^2$	$2p^6$	$3s^2$	$3p^6$		$4s^2$	
Sc	$1s^2$	$2s^2$	$2p^6$	$3s^2$	$3p^6$	$3d$	$4s^2$	
Ti	$1s^2$	$2s^2$	$2p^6$	$3s^2$	$3p^6$	$3d^2$	$4s^2$	
V	$1s^2$	$2s^2$	$2p^6$	$3s^2$	$3p^6$	$3d^3$	$4s^2$	
Cr	$1s^2$	$2s^2$	$2p^6$	$3s^2$	$3p^6$	$3d^5$	$4s$	
Mn	$1s^2$	$2s^2$	$2p^6$	$3s^2$	$3p^6$	$3d^5$	$4s^2$	
Fe	$1s^2$	$2s^2$	$2p^6$	$3s^2$	$3p^6$	$3d^6$	$4s^2$	
Co	$1s^2$	$2s^2$	$2p^6$	$3s^2$	$3p^6$	$3d^7$	$4s^2$	
Ni	$1s^2$	$2s^2$	$2p^6$	$3s^2$	$3p^6$	$3d^8$	$4s^2$	
Cu	$1s^2$	$2s^2$	$2p^6$	$3s^2$	$3p^6$	$3d^{10}$	$4s$	
Zn	$1s^2$	$2s^2$	$2p^6$	$3s^2$	$3p^6$	$3d^{10}$	$4s^2$	
Ga	$1s^2$	$2s^2$	$2p^6$	$3s^2$	$3p^6$	$3d^{10}$	$4s^2$	$4p$
Ge	$1s^2$	$2s^2$	$2p^6$	$3s^2$	$3p^6$	$3d^{10}$	$4s^2$	$4p^2$
As	$1s^2$	$2s^2$	$2p^6$	$3s^2$	$3p^6$	$3d^{10}$	$4s^2$	$4p^3$
Se	$1s^2$	$2s^2$	$2p^6$	$3s^2$	$3p^6$	$3d^{10}$	$4s^2$	$4p^4$
Br	$1s^2$	$2s^2$	$2p^6$	$3s^2$	$3p^6$	$3d^{10}$	$4s^2$	$4p^5$
Kr	$1s^2$	$2s^2$	$2p^6$	$3s^2$	$3p^6$	$3d^{10}$	$4s^2$	$4p^6$

We can now see that filling the $3d$ orbital gives rise to the familiar transition metal series; in exactly the same way the filling of the $4f$ orbital later in the Periodic Table gives rise to the lanthanide series. Table 3.3 shows a modern version of the Periodic Table; it is important to be able to see how the shape and structure of this table is derived from the quantum numbers we obtained from the Schrödinger equation, and Figure 3.15, which summarizes the order of energies of the orbitals. Much of the basic physics and chemistry of the elements can be rationalized by the consideration of the electron configurations of atoms, but it is not the task of this book to repeat this story. We must leave it to the books mentioned in the suggestions for further reading, while we follow the ideas of electron configuration, and of the energies and angular momenta of individual electrons to pursue further the energy levels and spectra of atoms as a whole.

Table 3.3   The Periodic Table

1	2	3	4	5	6	7	8	9	10	11	12	13	14	15	16	17	18
H 1																	He 2
Li 3	Be 4											B 5	C 6	N 7	O 8	F 9	Ne 10
Na 11	Mg 12											Al 13	Si 14	P 15	S 16	Cl 17	Ar 18
K 19	Ca 20	Sc 21	Tc 22	V 23	Cr 24	Mn 25	Fe 26	Co 27	Ni 28	Cu 29	Zn 30	Ga 31	Ge 32	As 33	Se 34	Br 35	Kr 36
Rb 37	Sr 38	Y 39	Zr 40	Nb 41	Mo 42	Tc 43	Ru 44	Rh 45	Pd 46	Ag 47	Cd 48	In 49	Sn 50	Sb 51	Te 52	I 53	Xe 54
Cs 55	Ba 56	La 57	Hf 72	Ta 73	W 74	Re 75	Os 76	Ir 77	Pt 78	Au 79	Hg 80	Tl 81	Pb 82	Bi 83	Po 84	At 85	Rn 86
Fr 87	Ra 88	Ac 89															

Lanthanides

La 57	Ce 58	Pr 59	Nd 60	Pm 61	Sm 62	Eu 63	Gd 64	Tb 65	Dg 66	Ho 67	Er 68	Tm 69	Yb 70	Lu 71

Actinides

Fk 89	Th 90	Pa 91	U 92	Np 93	Pu 94	Am 95	Cm 96	Bk 97	Cf 98	Es 99	Fm 100	Md 101	No 102	Lw 103

IV

The Energy Levels and Spectra
of Atoms

In this book we have pursued the logical course of discussing the electronic structure of atoms before any detailed consideration of their experimental spectra; historically however the study of the spectra led to the understanding of the electronic structure. Indeed atomic spectroscopy has played a crucial role in providing a firm theoretical basis for chemistry as well as physics. Before we can obtain a full description of electronic spectra of polyelectronic atoms, and so see the complete relationship between theory and experiments, we need two further pieces of theory.

We have so far considered in detail only the spectrum of the hydrogen atom, and have seen how wave mechanics and the four quantum numbers of the lone electron yield an energy diagram which provides a complete explanation of the main features of the spectrum. In the case of polyelectronic atoms, we are concerned not only with the energies and angular momenta of the individual electrons, which are given by the individual quantum numbers, but with the energy levels and angular momenta of the atom as a whole. Care is necessary when we try to find the resultant of the individual orbital and spin angular momenta. In the jargon of spectroscopists, we are concerned with the coupling of angular momenta.

A. Coupling of Angular Momenta

Angular momentum can be thought of as a vector quantity, that is it can be represented by an arrow, whose length represents its magnitude, and whose orientation represents the direction of the momentum (Figure 4.1). In this respect it is similar to velocity or force, and as for these classical mechanical quantities the resultant of two angular momenta v_1 and v_2 are given by a parallelogram of vectors: (Figure 4.2). Thus by trigonometry we can calculate both the magnitude and the direction of the resultant; it is intuitively obvious that if v_1 and v_2 represent for example the forces of two ropes pulling an object, then the object feels a pull in the direction of the resultant which is larger than either v_1 or v_2 individually.

In quantum mechanics the situation is in fact rather more simple,

Figure 4.1   Vector representation of angular momentum

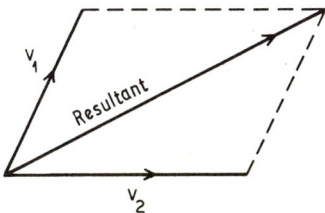

Figure 4.2   The resultant of
two vectors v_1 and v_2

and it is not necessary to use trigonometry to find the resultant. The vectors which we are trying to couple together are angula momenta, either orbital angular momentum, l, or spin angular momentum, s. As we have seen, these are quantized; l can only take integral values in units of $h/2\pi$ and s can only take the value ½. Not surprisingly, the resultant also has to be quantized; if this were not so, we should not observe the line spectra characteristic of all atoms.

B. Russell-Saunders Coupling

In light atoms the energy differences between the various energy levels of an atom arise primarily because of the different electrostatic energies of the various electronic configurations. There are also magnetic effects, which are due to the intrinsic magnetic moments of the electrons and the fields that they create by their motions. In Russell-Saunders coupling, in which we shall be primarily interested, the magnetic effects are treated as being of only secondary importance and are ignored as a first approximation.

We shall describe the actual coupling by a series of rules, and then illustrate these rules by considering some simple examples in detail. For Russell-Saunders coupling the procedure is as follows:

(i) The individual spin angular momenta of the electrons, s_i (each of which has a value ½), combine to give a resultant S which must be either integral or half-integral.

$$\therefore \Sigma s_i = S$$

For example, two spins of ½ could couple to give a resultant of $S = 1$ or $S = 0$. Thus coupling two spins of ½ need not only give a resultant of $S = 1$, as the ½ indicates the magnitude of the spin angular momentum only. If the spins have opposite directions, (m_s of the opposite sign), the resultant may have no spin, and $S = 0$.

(ii) The individual orbital angular momenta of the electrons, l_i (each of which may be 0, 1, 2, 3 ... for s, p, d, f electrons), combine to give a resultant L which must be quantized $L = 0, 1, 2, 3 \ldots$. Conventionally the value of the resultant orbital angular momentum is also indicated by roman letters, but for the resultant, capitals are used; thus

$$\Sigma l_i = L; \quad L = 0, 1, 2, 3 \ldots$$

referred to as $S, P, D, F \ldots$

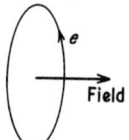

Figure 4.3   Magnetic field resulting from the motion of an electron

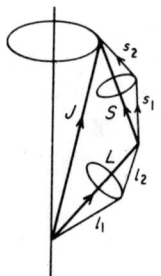

Figure 4.4   The Russell—Saunders coupling scheme

(iii) Now an orbiting electronic charge produces a magnetic field perpendicular to the plane of orbit (Figure 4.3) and so both the total orbital angular momentum L and the total spin angular momentum S have corresponding magnetic field vectors. As a result of this L and S can interact magnetically to give a resultant total angular momentum J; we are not surprised to learn that J is quantized in units of $h/2\pi$ having only integral or half-integral values. Figure 4.4 attempts to illustrate how the vectors combine. The possible values of the J quantum number are thus

$$J = (L + S), (L + S - 1) \ldots |L - S|.$$

C. Term Symbols

Long before there was any detailed understanding of atomic spectra, Balmer's work had made it clear that the frequency of any line in the hydrogen atom spectrum, ν, was a difference between two mathematical terms in the equation

$$E = h\nu = \frac{R}{n_1^2} - \frac{R}{n_2^2}$$

or

$$h\nu = T_1 - T_2$$

We know that these 'terms' represent energy levels. However, the old usage is so entrenched that the word 'term' is frequently used for an energy level. The name or label of an atomic energy level is referred to as a 'term symbol'.

Again for historical reasons the term symbols are written in a curious manner, which seems cumbersome, but which in practice is easy to remember and convenient to use. Each energy level (or term) of an atom is described by its quantum numbers L, S and J, which represent the orbital, spin and total angular momenta. The symbol summarizing these values is written

$$^{2S+1}L_J$$

The value of L is represented by the capital letter S, P, D etc. In the left superscript the value of S is coded in the form $2S + 1$. Thus if $S = 1$, the value of $(2S + 1)$ is 3, and so a 3 appears as a superscript in the term symbol. (Note that we are now using S in two distinct meanings in the same notation; S represents the total spin angular momentum, and is used to represent the value $L = 0$. This is an unfortunate consequence of the way in which the notation has developed; in practice it is rarely a source of confusion to the initiated, although this makes it no less irritating for the student.) The value of J appears as a right subscript.

Term symbols typical of those which we shall encounter are

$$^1S_0, \quad ^2S_{1/2} \quad \text{and} \quad ^3P_2.$$

These are read as 'singlet S nought', 'doublet S one half' and 'triplet P two', respectively.

D. Closed Electronic Shells

Before we go on to give some examples to clarify this curious system of nomenclature, we can introduce one important simplifying feature. When we have a complete shell of electrons, or even a complete sub-shell, such as s^2, p^6, d^{10} or f^{14}, then this set of electrons have resultant spin and orbital angular momenta equal to zero.

For a sub-shell, corresponding to the quantum numbers n and l, to be complete, every possible combination of the quantum numbers m_l and m_s must be represented by an electron. Thus for each value of m_l, there will be two electrons, one with spin $+\frac{1}{2}$ (α), and one with spin $-\frac{1}{2}$ (β). Every α spin will be counterbalanced exactly by a β spin, and the resultant spin angular momentum will be zero. The values of m_l will range from -1 to $+1$. There will be two electrons with values of $m_l = 0$, which have no orbital angular momentum, and then for each pair of electrons with a positive value of m_l, there will be another pair of electrons with the corresponding negative value of m_l. Again these electrons will have no resultant orbital angular momentum.

The effect of this simplification is firstly that all atoms whose ground states are closed shells, such as the noble gases

He $1s^2$

Ne $1s^2\ 2s^2\ 2p^6$

Ar $1s^2\ 2s^2\ 2p^6\ 3s^2\ 3p^6$

will have term symbols for their ground states $^1S_0$ (singlet S nought) — they have no resultant orbital angular momentum, no resultant spin angular momentum, and consequently no resultant total angular momentum.

A second consequence of this simplification is that even for atoms which have many electrons, it is rarely necessary to consider the angular momenta of more than a few electrons. All the electrons in a closed sub-shell contribute nothing to L, S or J and may be discounted. Thus the alkali metals, Li($1s^2\ 2s$), Na($1s^2\ 2s^2\ 2p^6\ 3s$) and K($1s^2\ 2s^2\ 2p^6\ 3s^2\ 3p^6\ 4s$) all reduce to one-electron problems, provided that we do not excite electrons out of the closed-shell core. Similarly, in the transition metals only the valence electrons, which are generally d electrons, are of importance in deciding the ground state term symbols.

E. Term Symbols of Some Simple Atoms

1. Hydrogen

The ground state of the hydrogen atom has the electron in a $1s$ orbital; as we are considering only one electron, there is no difference between l and L, and s and S, therefore

$$L = 0, \quad S = \frac{1}{2}, \quad \text{and} \quad J = \frac{1}{2}.$$

The term symbol for the ground state of the hydrogen atom is thus $^2S_{\frac{1}{2}}$. We can also easily write down the term symbols for some of the excited states of the hydrogen atom; for an electron in a $2s$, $3s$, $4s$ etc. orbital, it is always true that $L = 0$, $S = \frac{1}{2}$ and $J = \frac{1}{2}$, and therefore the term symbol for all these states is just $^2S_{\frac{1}{2}}$.

If we consider the $2p$ state of hydrogen, a slightly different situation arises. Again $S = \frac{1}{2}$, but now $L = 1$, and so there are two possible values of J, the total angular momentum, $J = \frac{1}{2}$ and $J = \frac{3}{2}$. There are therefore two states with term symbols $^2P_{1/2}$ and $^2P_{3/2}$. In the absence of magnetic effects these two states are degenerate; in practice, the spin and orbital angular momenta interact slightly. In the case of the hydrogen atom, the splitting between the $^2P_{1/2}$ and $^2P_{3/2}$ states is only 0.1 cm^{-1}, compared with a splitting between the ground $^2S_{1/2}$ and excited $^2P_{1/2}$ states of 83,000 cm^{-1}. In the same way, the $3p$, $4p$, $5p$ etc. states of the hydrogen atom each lead to two terms, with term symbols $^2P_{1/2}$ and $^2P_{3/2}$.

If we consider the term symbols from a d^1 configuration, we have $l = L = 2$, $s = S = \frac{1}{2}$, and therefore $J = \frac{5}{2}$ or $\frac{3}{2}$. Again there are two terms, $^2D_{3/2}$ and $^2D_{5/2}$ which are very close in energy. Remember that the smallest allowed value of J is $(L - S)$, and so $J = \frac{1}{2}$ is not a permitted value of the total angular momentum quantum number.

We can now re-draw Figure 3.13, labelling the states with the correct term symbols, and this is shown in Figure 4.5.

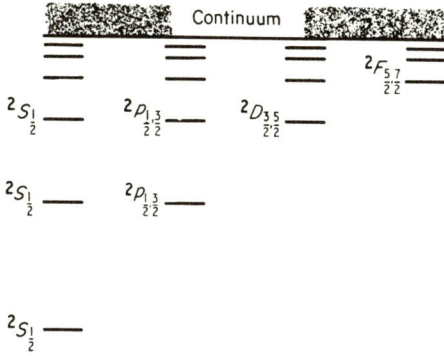

Figure 4.5   Term symbols for the hydrogen atom energy levels

2. Lithium

The ground state of the lithium atom has the configuration $1s^2\,2s$; the $1s$ shell is complete, and therefore makes no contribution to the resultant angular momenta, so that the problem reduces to a one-electron case. Thus for the ground state,

$$L = 0 \quad S = \frac{1}{2} \quad J = \frac{1}{2},$$

and the term symbol is $^2S_{1/2}$. If we consider the excited states $1s^2\,3s$, $1s^2\,4s$ etc., again we obtain a series of $^2S_{1/2}$ states, each having a higher energy than the previous one. Thus we shall obtain an energy level diagram for the $^2S_{1/2}$ states like that shown in Figure 4.6. It is important to note that there is now no simple relationship between the energy levels, as there is in the corresponding hydrogen atom case.

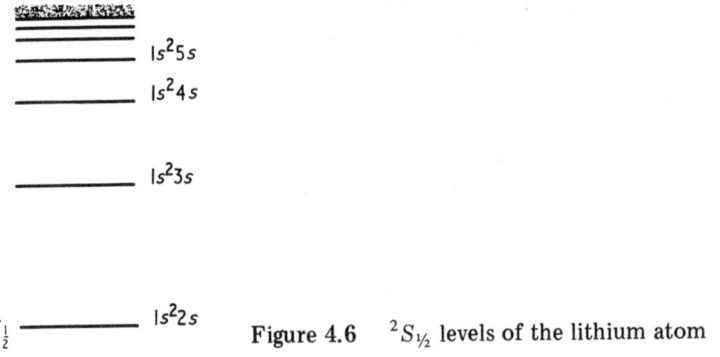

$^2S_{\frac{1}{2}}$ ———— $1s^2 2s$ **Figure 4.6** $^2S_{\frac{1}{2}}$ levels of the lithium atom

There is also a series of excited states $1s^2\, 2p$, $1s^2\, 3p$, $1s^2\, 4p$, and just as in the hydrogen atom case, these each give rise to two terms, $^2P_{\frac{1}{2}}$ and $^2P_{\frac{3}{2}}$. In the absence of magnetic effects, these would be degenerate; in fact, there is a small splitting between them. Figure 4.7 shows the energy level diagram for the $1s^2\, np$ states. It is important to remember that these states are *not* degenerate with the $1s^2\, ns$ states; in fact they lie a little higher. In the hydrogen atom, the ns and np states are virtually degenerate. Precisely similar considerations govern the series of levels corresponding to the configurations $1s^2\, 3d$, $1s^2\, 4d$, etc; here $L = 2$, $S = \frac{1}{2}$, and $J = \frac{3}{2}$ or $\frac{5}{2}$. The complete energy level diagram for the lithium atom is shown in Figure 4.8. Note that configurations in which an electron is excited from the $1s^2$ core are of too high an energy to appear in this diagram.

$1s^2 5p$
$1s^2 4p$

$1s^2 3p$

$^2P_{\frac{3}{2}}$
$^2P_{\frac{1}{2}}$ $1s^2 2p$

Figure 4.7 ^2P levels of the lithium atom (not to scale)

3. Helium

The helium ground state has the closed shell configuration $1s^2$ and so as we have seen has the term symbol $^1S_0$.

The simplest excited state has the configuration $1s\, 2s$, and the problem is now more complicated as there are two electrons to consider.

$$l_1 = 0 \qquad l_2 = 0 \qquad \therefore L = 0$$
$$s_1 = \frac{1}{2} \qquad s_2 = \frac{1}{2} \qquad \therefore S = 0 \text{ or } 1.$$
$$\therefore J = 0 \text{ or } 1.$$

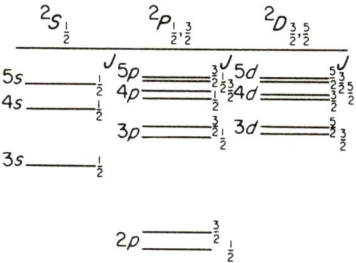

Figure 4.8 Energy levels of the lithium atom (not to scale)

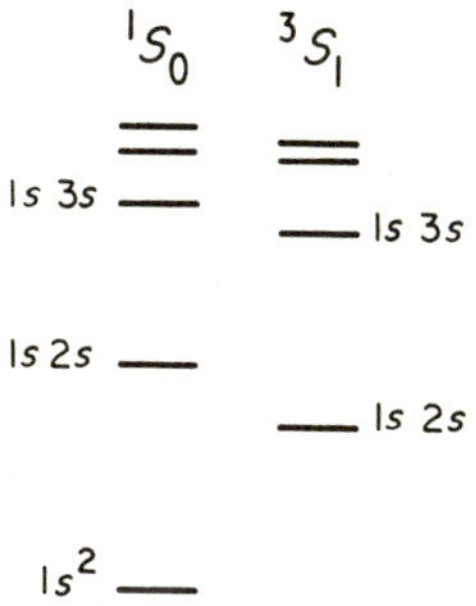

Figure 4.9 S levels of helium

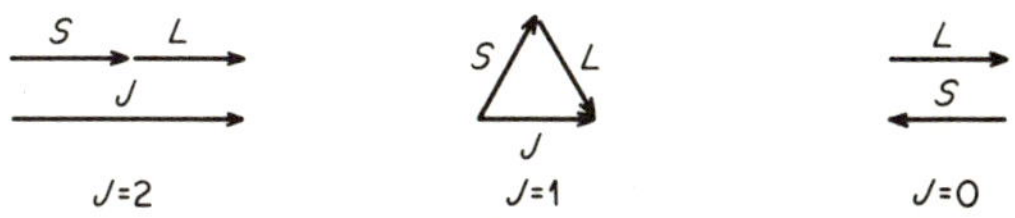

Figure 4.10 Coupling of L and S for ^3P states of He

There are thus two terms, $^1S_0$ and $^3S_1$ and these are not degenerate, even in the absence of magnetic effects. The same argument holds for the excited configurations $1s\,3s$, $1s\,4s$, etc., and so we can draw part of the energy level diagram as two separate energy ladders (Figure 4.9).

The states with the configurations $1s\,np$ have;

$$l_1 = 0 \quad l_2 = 1 \quad \therefore L = 1$$
$$s_1 = \tfrac{1}{2} \quad s_2 = \tfrac{1}{2} \quad \therefore S = 0 \text{ or } 1$$
$$\therefore J = 1 \text{ or } 0, 1, 2.$$

If $S = 0$, then $J = 1$, and we have a $^1P_1$ term; but if $S = 1$, then there are three possible values of the J quantum number, arising from $S = 1$ and $L = 1$ by vector addition (Figure 4.10). There are thus three triplet

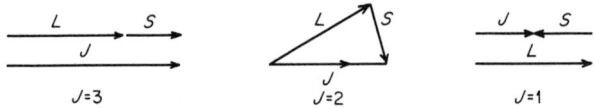

Figure 4.11   Coupling of L and S for ^3D states of He

terms, $^3P_2$ $^3P_1$ and $^3P_0$ and the splitting between these is very small; but again, the splitting between these terms and the $^1P_1$ term is caused by electrostatic interactions rather than magnetic effects and is much larger.

For the $1s\,nd$ states, we have a $^1D_2$ state, and three triplet components, $^3D_3$ $^3D_2$ and $^3D_1$. The coupling for the triplet states is shown in Figure 4.11.

F. Selection Rules

We have now seen how we can understand the energy levels of simple atoms, and how we can obtain term symbols to describe the coupling of angular momenta for each energy level. We are still not able to predict the appearance of the absorption spectra of these simple atoms, however, because we have not yet taken account of the so-called 'selection rules'. It is found experimentally that transitions between all possible combinations of states are not observed; only a limited number of transitions are allowed and these are governed by the selection rules.

Before we state the selection rules and consider their effect on observed spectra, we must consider how they arise theoretically. This illustrates how powerful a technique quantum mechanics is even at a practical level. If a wave function is available for a particular electronic state, then it is in principle possible to calculate any observable property of that state. The property which presently concerns us is the intensity of a spectral transition between two states which we shall label I and II. It may be shown that intensity is proportional to an integral over all space which involves the wave functions of the two states, Ψ_I and Ψ_{II} together with a vector $\mathbf{R}$.

$$\text{Intensity} \propto \int \Psi_I \mathbf{R} \Psi_{II} d\tau \tag{4.1}$$

We do not need to consider this integral in great detail; consideration of some simple properties of the integral will lead to an appreciation of selection rules. The integral is a definite integral over symmetric limits, that is it is of the type

$$\int_{-\alpha}^{\alpha} y dx$$

where y is a function of x. Such integrals may be divided into two

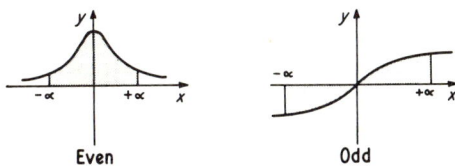

Figure 4.12   Definite   integrals   over symmetric limits

types, shown in Figure 4.12. The shaded areas represent the definite integral, and it can be shown that in the case labelled 'odd', the negative and positive areas exactly cancel, so that definite integrals of odd functions over symmetric limits have the value zero. This is important to our consideration of selection rules, for if we can show that the function in the integral for calculating intensity (equation 4.1) is odd, then the intensity will be zero and the transition will not appear in the spectrum.

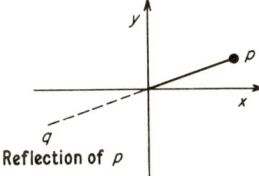

Figure 4.13   Reflection at the origin

The functions Ψ_I and Ψ_{II} are even or odd depending on whether they are s, p, d or f. The even functions are unchanged if their coordinates are reversed (equivalent to a reflection at the origin, as in Figure 4.13), while odd functions retain the same magnitude, but change sign on reflection. Thus s and d functions are even, and p and f functions are odd, as may be seen from Figure 4.14. The vector $\mathbf{R}$ is an odd quantity, as can be seen from Figure 4.15, since on reversing $x \rightarrow -x$, $y \rightarrow -y$ $z \rightarrow -z$, so $\mathbf{R} \rightarrow -\mathbf{R}$.

The rules for obtaining the symmetry of the product of two

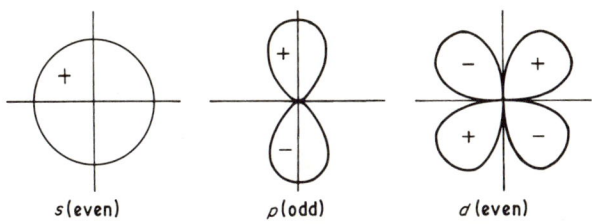

Figure 4.14   Symmetry properties of atomic orbitals

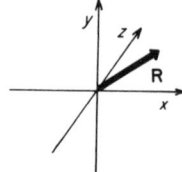

Figure 4.15   The vector R in three dimensions

functions are simple:

even x even = odd x odd = even

odd x even = even x odd = odd

As the product $(\Psi_I R \Psi_{II})$ must be even for the intensity to be non-zero, then one of Ψ_I and Ψ_{II} must be odd, and the other even. Considerations of this type give rise to the selection rules for atomic transitions which are summarized below.

(i) There is no selection rule governing the change in value of n, the principal quantum number, which any electron may undergo. Thus in the hydrogen atom, the transitions $1s - 2p$, $1s - 3p$, $1s - 4p$, etc. are all allowed.

(ii) The selection rule for the total orbital angular momentum is

$$\Delta L = \pm 1$$

Thus transitions such as $^1S - {}^1P$ and $^2D - {}^2P$ are allowed, but transitions such as $^3D - {}^3S$, for which $\Delta L = -2$, are forbidden, and do not appear in the spectrum. An alternative way of rationalizing this is in terms of conservation of angular momentum; photons have one unit of angular momentum.

(iii) The selection rule for the spin angular momentum is

$$\Delta S = 0$$

Thus transitions such as $^2S - {}^2P$ and $^3D - {}^3F$ are allowed, but transitions such as $^1S - {}^3P$ are forbidden.

(iv) The selection rule for the total angular momentum, J, is

$$\Delta J = 0 \text{ or } \pm 1$$

Thus transitions such as $^2P_{1/2} - {}^2D_{3/2}$ and $^2P_{3/2} - {}^2D_{3/2}$ are allowed, but transitions such as $^2P_{1/2} - {}^2D_{5/2}$ for which $\Delta J = 2$, are forbidden.

We can now see how important it is to obtain the correct term symbols for electronic states, for the term symbols govern whether a particular transition is allowed or forbidden. We can now return to the discussion of the electronic states of lithium and helium, and predict the experimental spectrum from our knowledge of the positions of the states and of the relevant selection rules.

It is important to realize that the section on lithium applies equally to the other alkali metals Na and K; for the electron configurations of the alkali metals are all of the type [inert gas core] ns^1 and, as we have seen, the inert gas cores play no role in the optical properties of atoms. Thus in each case, the problem is similar and is that of a single s electron outside an inert gas core. Similarly the He spectrum may act as a model for understanding the spectra of the alkaline earth metals, such as Mg([Ne core] $3s^2$) and Ca([Ar core] $4s^2$), and also elements such as Zn([Ar core] $4s^2 3d^{10}$), where the $3d$ sub-shell is filled and does not materially affect the optical spectrum.

G. The Spectra of Li and He

1. Lithium

As we have seen, the ground electronic state for the Li atom has the configuration $1s^2 2s$, and the term symbols $^2S_{1/2}$. There is a series of excited states of the type $1s^2 ns$, and these all have term symbol $^2S_{1/2}$. We do not see transitions between these states and the ground state as they all have the same value of the L quantum number and the transition is therefore forbidden ($\Delta L = 0$).

However, it is possible to observe transitions between the ground state and the excited states of the type $1s^2 2p$. These excited states have term symbols $^2P_{1/2}$ and $^2P_{3/2}$ and therefore transitions to both levels are allowed ($\Delta S = 0$, $\Delta L = 1$, $\Delta J = 0$ or 1). As we have seen, because of magnetic effects the $^2P_{1/2}$ and $^2P_{3/2}$ levels are not exactly degenerate, and this gives rise to two closely-spaced lines in the spectrum, called a doublet (Figure 4.16). In exactly the same way, transitions to all excited states of the type $1s^2 np$ are allowed, and so the spectrum consists of a set of doublets, as shown in Figure 4.17.

States coming from configurations of the type $1s^2 nd$ are $^2D_{3/2}$ and $^2D_{5/2}$ and transitions to these states from the ground state are not allowed as $\Delta L = 2$. Similarly transitions from the ground state to ^2F states, from the configuration $1s^2 nf$, are forbidden ($\Delta L = 3$).

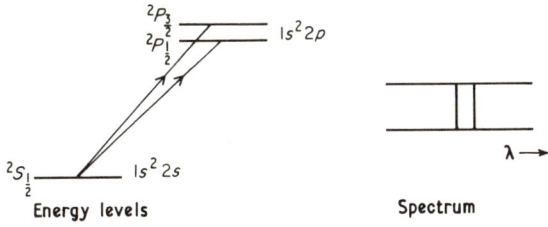

Figure 4.16   The lithium doublet transition

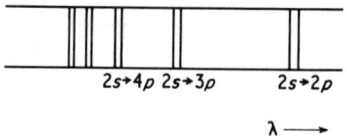

Figure 4.17 Schematic representation of part of the lithium spectrum

We have so far considered only transitions from the ground state. These transitions constitute the absorption spectrum at relatively low temperatures; but in the emission spectrum, observed at higher temperatures, we may observe transitions which do not involve the ground state. Figure 4.18 shows some of the transitions which may be observed in the emission spectrum.

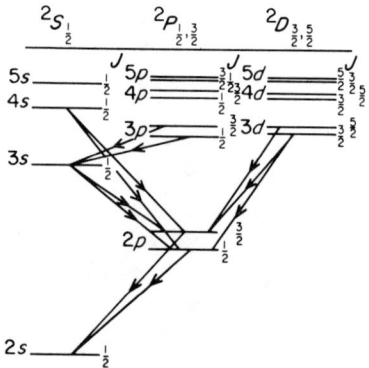

Figure 4.18 Some of the transitions observed in the lithium emission spectrum

There are four important series of lines which are known as:

the SHARP series	$ns - mp$
the PRINCIPAL series	$np - ms$
the DIFFUSE series	$nd - mp$
the FUNDAMENTAL series	$nf - md$

The names were bestowed on the series before there was any detailed understanding of their natures, but they have been retained to this day in the nomenclature of s, p, d and f orbitals.

2. Helium

We have seen that excited states of the He atom may be divided into two groups, those with $S = 1$ (triplet states) and those with $S = 0$

(singlet states). This is an important distinction to make, because transitions in which $\Delta S \neq 0$ are forbidden, and so we need concern ourselves only with transitions between two singlet states and between two triplet states.

The ground state of the He atom is $1s^2$, $^1S_0$. Thus in the absorption spectrum, where we are exciting atoms from the ground state, we observe only singlet–singlet transitions. Clearly as $\Delta L = \pm 1$, we can observe transitions only to $^1P_1$ states, which arise from the configurations $1s\, np$. Note that there is only one possible value of J if $S = 0$ and $L = 1$, and so the spectrum appears as a set of single lines, as shown in Figure 4.19.

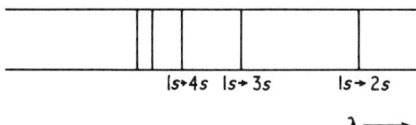

Figure 4.19 Schematic representation of part of the helium absorption spectrum

In emission, the situation is a little more complicated. In the singlets, not only $S - P$ transitions, but also $P - D$ transitions, may be observed; the situation is rather similar to that in Li (Figure 4.18), but in each transition only a single line is observed. This is shown in Figure 4.20.

It is also possible to obtain triplet–triplet transitions in emission

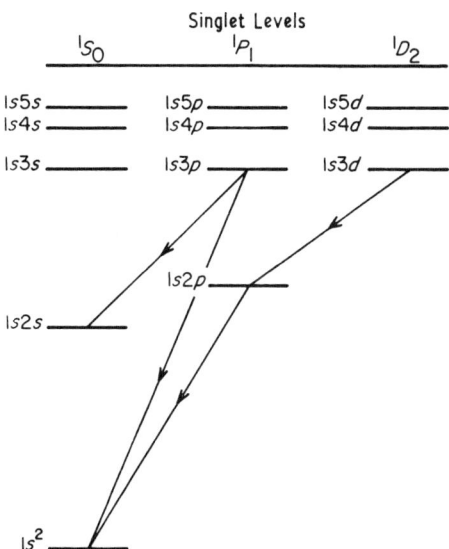

Figure 4.20 Some of the transitions observed in the helium singlet system

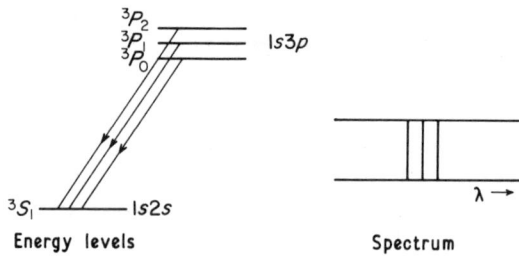

Figure 4.21   A triplet transition in helium

spectra of helium. The general scheme is rather similar to that for the singlet states, but there is an important difference in the appearance of the lines. A typical triplet transition would be from the state $1s\,3p$ to the state $1s\,2s$. For the upper state, $S = 1$ and $L = 1$, and therefore $J = 0$, 1 or 2. There are therefore three levels which in reality are not quite degenerate. For the lower state $S = 1$ and $L = 0$, and therefore $J = 1$. The selection rules show that transitions are allowed to the lower state from each of the upper levels, and therefore the spectrum appears as three narrowly-spaced lines; this is illustrated in Figure 4.21. The energy level diagram for the triplet states of He is given in Figure 4.22; each allowed transition gives rise to three narrowly spaced lines, as in Figure 4.21.

One problem which can cause confusion is how it is possible to observe triplet–triplet transitions if we begin with a sample of helium in its singlet ground state. No singlet–triplet transitions are observed, so the triplet states cannot be populated directly; a simple mechanism for populating the triplet states is via the He^{+} ion. If an electron is excited into the continuum, an He^{+} ion is formed. If the electron is

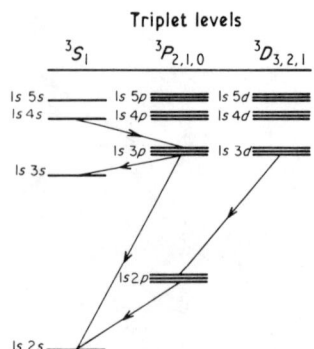

Figure 4.22   Some of the transitions observed in the helium triplet system

subsequently recaptured, then an atom either in a singlet or a triplet state can be formed; in fact, a triplet state is three times as likely to be produced as a singlet state.

H. The He Spectrum as a Model for Photochemistry

Sometimes the study of atomic spectra may seem to be somewhat remote from mainstream chemistry, but the He spectrum provides a typical example of how a simple atomic system can provide a model for large areas of chemistry of current interest.

Most molecules which exist naturally are stable and not very reactive because all their electrons have paired spins; they are closed-shell molecules. Excitation by light to an excited state will split one electron pair, leaving the other electron pairs substantially unaffected. The excited electron pair is thus similar to the electron pair in the He atom. The molecules have ground states which are singlets, and are generally labelled S_0 (to add to possible confusion over the meanings of S in spectroscopy). Excitation, as in helium, can lead to either a singlet state, labelled S_1, or a triplet state, labelled T_1, although of course the triplet cannot be reached by direct photon absorption, as ΔS would then be 1. The energy levels are shown in Figure 4.23. The triplet level lies below the corresponding singlet level, as by Hund's rules given earlier a state with two electrons with parallel spins is more favourable than the corresponding state where the spins are anti-parallel.

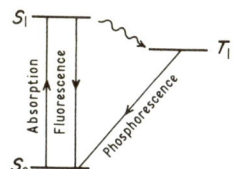

Figure 4.23 Photochemical processes in an organic molecule

If a molecule is excited to the state S_1 then it may be able to radiate its energy, falling back to the ground state. This process is called *fluorescence* and takes times of the order of 10^{-8} seconds. As a result of spin-orbit coupling, which we shall discuss in the next section, it is also possible for the triplet state T_1 to be populated from the excited singlet state by *intersystem crossing*. This is shown as a wavy arrow in Figure 4.23. The transition from the triplet state to the ground state is in principle forbidden, as $\Delta S = -1$. In practice it is found that this selection rule does not hold absolutely; the transition $T_1 - S_0$ does take place, though very slowly, taking times of the order of a few seconds to occur. This delayed emission of light is called *phosphorescence*.

The chemical importance of light is largely based on the long lifetime of the triplet state. Molecules in this state have a large amount of energy, and also two unpaired electrons; they can take part in very important and striking chemical reactions. During the lifetime of the triplet state, which is long on the molecular time-scale, the molecule can collide with many other molecules or undergo geometrical rearrangements before it can lose its energy in phosphorescence.

I. Spin-orbit Coupling

We have seen that in the spectra of the hydrogen atom and the alkali metals, a splitting is observable in the ^2P states, and that this splitting is caused by a magnetic effect. The effect is called spin-orbit coupling, and is caused because the spinning electron acts as a tiny magnet which interacts with the field created by its motion around the nucleus.

For a state with a wave function Ψ the splitting depends approximately on the size of the integral

$$\int \Psi \frac{Z}{r^3} \Psi d\tau$$

where Z is the effective charge on the nucleus and r is the distance of the electron from the nucleus. The average distance of the electron from the nucleus is roughly inversely proportional to Z, and so the splitting is roughly proportional to Z^4. This is illustrated in Figure 4.24, where the splittings of the lowest ^2P states of the alkali metals are shown.

This high dependence of the spin-orbit splitting on Z may be appreciated better if we realize that the magnetic interaction of a charge e rotating around a nucleus of charge Z is exactly equivalent to

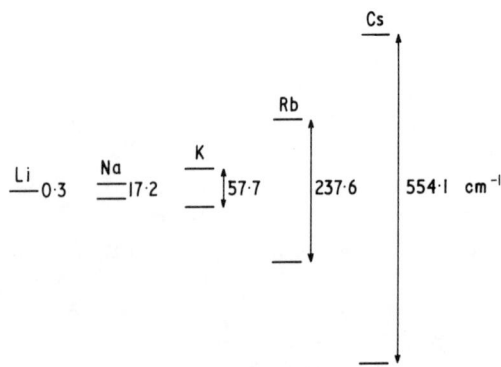

Figure 4.24   Splittings of the $^2P_{3/2}$ and $^2P_{1/2}$ levels of the alkali metals (cm^{-1})

the nuclear charge Z rotating round the charge e. Clearly then the magnetic effect is bigger the larger the value of Z.

It is spin-orbit coupling which allows an atom to undergo a transition from a singlet to a triplet state — a transition which is forbidden by the selection rules. It is often said that the coupling of spin and orbital angular momenta mixes the multiplicities of terms, and that singlet terms for example acquire some triplet characteristics. It follows that L and S are thus not 'good' quantum numbers.

J. jj Coupling

In this chapter all the energy level diagrams which we have drawn have been based on the Russell-Saunders coupling scheme, in which electrostatic interactions between electrons have dominated term differences and magnetic effects have been ignored or treated only as small perturbations. In *jj* coupling, magnetic effects dominate the term differences and electrostatic effects play a more minor role.

In *jj* coupling, the rules for coupling the angular momenta are as follows;

(i) For each electron i, the orbital and spin angular momenta l_i and s_i combine to give the total angular momentum for the electron j_i.

(ii) The individual j_i then couple to give a resultant J, the total angular momentum of the atom.

The method is best illustrated by an example, and we shall consider the excited states of He, with one s and one p electron.

$$l_1 = 0 \quad s_1 = \tfrac{1}{2} \quad \therefore j_1 = \tfrac{1}{2}$$
$$l_2 = 1 \quad s_2 = \tfrac{1}{2} \quad \therefore j_2 = \tfrac{1}{2} \text{ or } \tfrac{3}{2}$$

We now have two states, which are labelled $(\tfrac{1}{2}, \tfrac{3}{2})$ and $(\tfrac{1}{2}, \tfrac{1}{2})$, the symbols indicating the values of j_1 and j_2. The relative energies of the terms from an sp configuration in helium are shown in Figure 4.25.

It is important to see that the number of terms from the configuration sp is the same whether we treat it by Russell-Saunders or

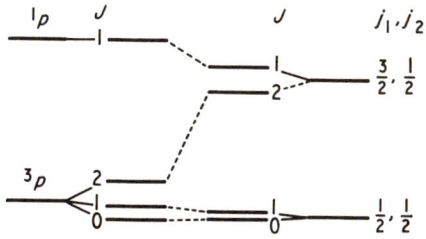

Figure 4.25 The transition from Russell-Saunders to *jj* coupling

jj coupling. In fact these two methods represent two extremes; in one case, magnetic effects are much greater than electrostatic effects, and in the other case, the reverse is true. Reality always lies somewhere between these two extremes, and we must use whichever coupling scheme is more appropriate to a particular problem; in general, light atoms are best treated by Russell—Saunders coupling and very heavy atoms by *jj* coupling.

In the next chapter we shall go on to investigate the effects of *external* magnetic fields on energy levels and observed spectra.

V

The Effects of Magnetic and
Electric Fields

If we look at the spectrum of the sodium atom by observing the light from a sodium lamp, we observe a discrete line spectrum, whose strongest features are the two D lines in the yellow region. However, if the lamp is placed in a magnetic field, the lines split, and the magnitude of the splitting is roughly proportional to the strength of the magnetic field. The magnetic field has removed degeneracies, and the actual levels have been split. The details of these splittings are connected, perhaps not surprisingly, with the quantum number m_l, about which we have said relatively little. This quantum number is commonly referred to as the space or magnetic quantum number, and it refers of course to a single electron. We shall also be concerned in this chapter with the quantum number M_J, which is the quantized component of the total angular momentum of an atom, J.

A. The Normal Zeeman Effect

In order to understand the effect of a magnetic field on an atom, we can begin by calculating the magnetic moment caused by the orbital motion of an electron. Classical physics gives us the relationship

$$\mu = -\frac{pe}{2m}$$

where p is the angular momentum and μ is the resultant magnetic moment. We know that in an atom the orbital angular momentum is quantized and that

$$P = Lh/2\pi$$

[more correctly $\sqrt{L(L+1)}\, h/2\pi$] and so

$$\mu = -L\frac{he}{4\pi m} \tag{5.1}$$

$$= -L\beta$$

where β is the Bohr magneton.

Now let us consider the effect of a magnetic field on an atom in a ^1P state. In this case, $S = 0$ and $L = 1$, so $J = 1$ and the magnetic moment arises only from the orbital angular momentum. When the magnetic field is applied, the vector J, representing the total angular momentum, begins to precess around the field; this effect is known as Larmor precession. The energy of interaction, ΔE, between the magnetic moment and the field is given by

$$\Delta E = \mu_B \times B$$

where μ_B is the component of μ in the direction of the field, B. From equation (5.1) we can see that

$$\mu_B = - M_J \frac{he}{4\pi m}$$

$$\therefore \Delta E = - M_J \frac{heB}{4\pi m} \qquad (5.2)$$

The term $(eB/4\pi m)$ has the units $\sec^{-1}$, and may be shown to be equal to the frequency of the Larmor precession.

From equation (5.2) we can deduce that there are three possible values of ΔE corresponding to the possible values of M_J of -1, 0 and 1. The ^1P term is thus split into three levels, with a splitting of βB. Now if we consider an atom in a ^1S state, we can see that as $S = L = J = 0$, the atom has no angular momentum, no magnetic moment and is consequently unaffected by a magnetic field. We can therefore predict the effect of a magnetic field on a $^1P - {}^1S$ transition; this is illustrated in Figure 5.1. The upper ^1P term is split into three levels, whereas the lower ^1S term is unaffected by the field. All three possible transitions are allowed and so the single line is split to become a triplet, with a splitting proportional to the field strength.

We can similarly predict the effect of a magnetic field on a $^1D - {}^1P$ transition; the ^1D state, with $S = 0$ and $L = J = 2$, is split into five levels, corresponding to five possible values of the M_J quantum number. This is shown in Figure 5.2. There are two important features of Figure 5.2; firstly, the splittings in the upper and lower states are equal, so each

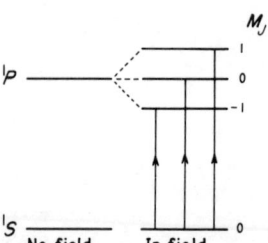

Figure 5.1 Effect of a magnetic field on a $^1P - {}^1S$ transition

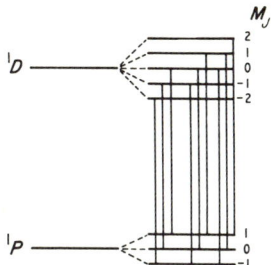

$$M_J$$

Figure 5.2   Zeeman effect for a $^1D - {}^1P$ transition

observed line corresponds to a particular value of ΔM_J, and includes transitions from each of the levels of the ^1P term; secondly, only transitions which satisfy the selection rule

$$\Delta M_J = 0, \pm 1$$

are allowed, and so the transition still has the appearance of a triplet, with the splitting proportional to the field strength.

It is clear that the effect of a magnetic field on any transition between singlet states is to split the original single line into three components. This is known as the normal Zeeman effect and applies only to singlet—singlet transitions.

B. The Anomalous Zeeman Effect

The anomalous Zeeman effect is in fact much more common than the normal Zeeman effect, and occurs when a magnetic field is applied to an atom in a state with $S \neq 0$, that is any state other than a singlet. The difference between cases in which $S = 0$ and those in which $S \neq 0$ is caused by the magnitude of the magnetic moment associated with the spin angular momentum of the atom.

We showed by classical analogy that the magnetic moment due to the orbital angular momentum of a single electron was given by

$$\mu = -l\beta$$

where β is the Bohr magneton (equation 5.1). The spin angular momentum also gives rise to a magnetic moment, but its magnitude is given by

$$\mu = -gs\beta \quad \text{where } g = 2.$$

The figure 2 arises from quantum mechanical details which differ from classical physics; it has been called the gyromagnetic anomaly. We do not need to understand the origin of this factor in any detail; we need to know only that as a result of the inclusion of this factor, the total magnetic moment of an atom is given by

$$\mu = -Jg\beta$$

where

$$g = \frac{3J(J+1) + S(S+1) - L(L+1)}{2J(J+1)} \tag{5.3}$$

The energy of interaction of the magnetic moment with the field is thus given by

$$\begin{aligned}
\Delta E &= -\mu_B \times B \\
&= -M_J g\beta B \\
&= -M_J g \frac{heB}{4\pi m}
\end{aligned} \tag{5.4}$$

If this expression is compared with that obtained in equation (5.2) for the singlet case, it can be seen that the general pattern of splitting levels with different values of M_J remains, but that the magnitude of the splitting now depends on the individual values of L and S, which are reflected in the value of g, the so-called Landé g-factor. We can see the effect of the Landé g-factor by considering a particular example, the D-lines in sodium.

C. The Sodium D-lines

The energy levels responsible for the prominent D-lines in the sodium spectrum are shown in Figure 5.3. We may use equation (5.3) to calculate the Landé g-factors for each level.

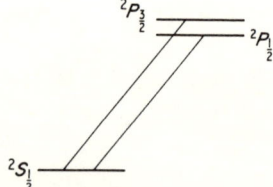

Figure 5.3 Transition responsible for the sodium D lines

For the $^2S_{1/2}$ level, $L = 0$, $J = \frac{1}{2}$ $S = \frac{1}{2}$ $\therefore g = 2$

for the $^2P_{1/2}$ level, $L = 1$, $J = \frac{1}{2}$ $S = \frac{1}{2}$ $\therefore g = \frac{2}{3}$

for the $^2P_{3/2}$ level, $L = 1$, $J = \frac{3}{2}$ $S = \frac{1}{2}$ $\therefore g = \frac{4}{3}$

Each of the levels is therefore split by the magnetic field according to the equation $\Delta E = -M_J g\beta B$, and this is shown in Figure 5.4. The spectrum does not now contain only three lines, because the splittings

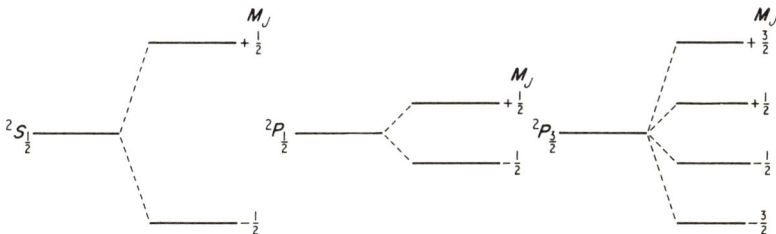

Figure 5.4   Anomalous splitting of levels in a magnetic field

in the upper and lower states are no longer identical. With the use of the selection rule

$$\Delta M_J = 0, \pm 1$$

it may be shown that there are ten possible transitions which are allowed; two transitions, $-\frac{1}{2} \rightarrow +\frac{3}{2}$ and $+\frac{1}{2} \rightarrow -\frac{3}{2}$ are forbidden. The $^2P_{\frac{1}{2}} - {}^2S_{\frac{1}{2}}$ transition is thus split into four lines, as shown in Figure 5.5, and the $^2P_{\frac{3}{2}} - {}^2S_{\frac{1}{2}}$ transition is split into six lines, as illustrated in Figure 5.6.

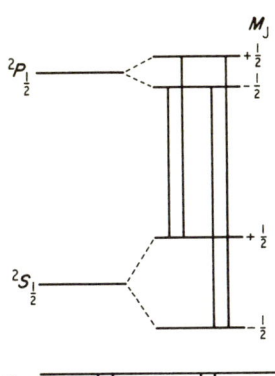

Spectrum

Figure 5.5   Anomalous   Zeeman   effect   for   a $^2P_{\frac{1}{2}} - {}^2S_{\frac{1}{2}}$ transition

D. The Paschen-Bach Effect

If the magnetic field which is applied to an atom is very strong, the spin and orbital angular momenta S and L uncouple from each other and each couples directly to the applied field. The momenta, and therefore their magnetic moments, precess independently about the magnetic field. Now an electromagnetic wave can interact only with L, and not with S, and so under these circumstances,

$$\Delta S = 0$$

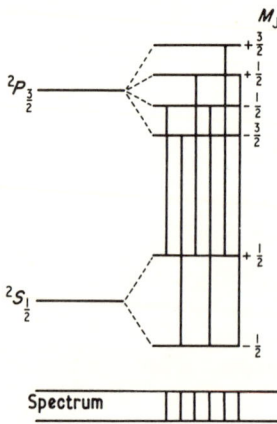

M_J

$^2P_{\frac{3}{2}}$

$+\frac{3}{2}$
$+\frac{1}{2}$
$-\frac{1}{2}$
$-\frac{3}{2}$

$+\frac{1}{2}$

$^2S_{\frac{1}{2}}$

$-\frac{1}{2}$

Spectrum

Figure 5.6 Anomalous Zeeman effect for a $^2P_{\frac{3}{2}} - {}^2S_{\frac{1}{2}}$ transition

It is therefore not possible to observe the effects associated with the anomalous magnetic moment associated with electron spin, and so the normal Zeeman effect is observed. This transition from the anomalous Zeeman effect to the normal Zeeman effect at high magnetic fields is called the Paschen-Bach effect.

E. The Stark Effect

Similar in some respects to the Zeeman effect is the Stark effect which is the splitting of spectral lines caused not by a magnetic field, but by an electric field.

The interaction of an electric field with an atom induces an electric dipole moment, the electrons being attracted in one direction and the nucleus in the other. The induced dipole μ_z is proportional to the applied field E_z, the proportionality constant being called the polarizability of the atom, α.

$$\mu_z = \alpha E_z$$

This dipole moment will precess round the field with quantized components dictated by the possible values of M_j. The energy of interaction of the field with the atom is proportional to the product of the field strength and the induced dipole moment, and thus is proportional to the square of the field strength. Reversing the direction of the field therefore has no effect on the interaction energy, and therefore levels with the same value of M_J, but opposite sign, have the same energy in the presence of a field. A further difference between the Stark and Zeeman effects is that the splitting of spectral lines is not symmetrical about the field-free positions. The effect of an electric field on the sodium D lines is illustrated in Figure 5.7.

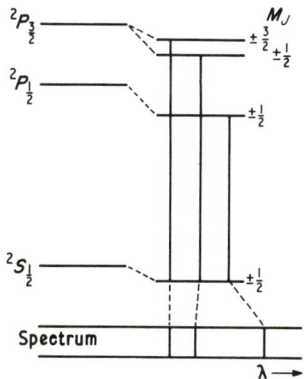

Spectrum

$\lambda \longrightarrow$

Figure 5.7   Stark effect on the sodium D lines

F. Crystal Field Theory

There is one extremely important example of the Stark effect in chemistry. The magnitude of electric fields which can be produced using charged plates is sufficient to produce only small shifts and splittings of spectral lines. But much stronger fields are encountered when the field is produced on the molecular level, as in a transition metal complex such as the ion $Fe(CN)_6^{3-}$ (Figure 5.8). The effect on the central atom is now much larger, as the negatively charged CN^- ions are only a few Ångstroms away. The resultant splittings are now so vast that the separations between components are energies that correspond to the visible part of the spectrum. Transitions may be induced between these components, which is one of the reasons why so many transition metal compounds and complexes are highly coloured.

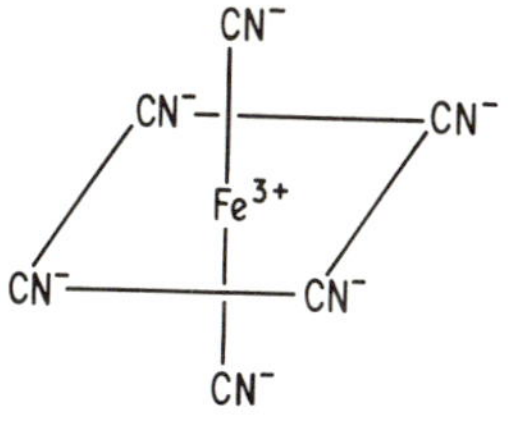

Figure 5.8   The $Fe(CN)_6^{3-}$ ion

G. Statistical Weights

Figure 5.9 illustrates the effect on the energy levels of the sodium atom of gradually reducing an applied magnetic field to zero. We can see that as the field intensity decreases to zero, the $^2P_{3/2}$ term consists of four degenerate levels, and the $^2P_{1/2}$ and $^2S_{1/2}$ terms each consist of two degenerate levels. It has never been found possible to split these levels, (except by nuclear hyperfine effects, to be discussed in Chapter 6) and

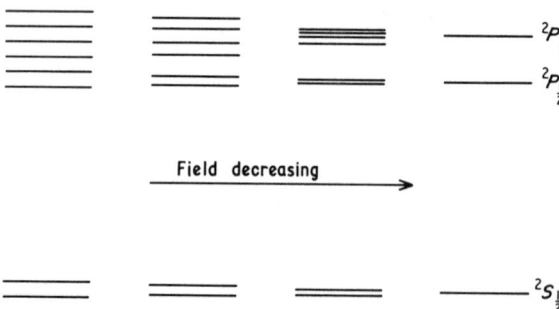

Figure 5.9 Effect of reducing a magnetic field on ^2P and ^2S levels

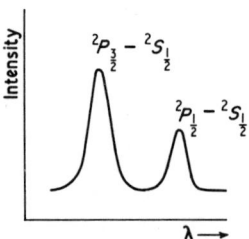

Figure 5.10 The intensities of the sodium D lines

they must therefore be regarded as simple. The number of levels in a particular term is equal to $(2J + 1)$, that is the number of possible values of M_J. It is known as the *statistical weight* of the term.

The importance of statistical weights derives from the suggestion that all levels have the same *a priori* probability, that is they all appear equally often under the same conditions. This suggestion has been verified experimentally, and so, as the statistical weight of a $^2P_{3/2}$ term is 4, and that of a $^2P_{1/2}$ term is 2, it is found that the intensities of the D lines in sodium are in the ratio 2:1 (Figure 5.10). In general, for the states with J values J_1 and J_2, the probabilities of finding an atom in those states are in the ratio $(2J_1 + 1):(2J_2 + 1)$, provided that the states have roughly equal energies.

H. The Stern-Gerlach Experiment

The statistical weight of energy levels is shown up in a dramatic illustration of space quantization which was first performed by Stern and Gerlach. In this experiment a beam of metal atoms is obtained by heating a sample of the metal in a furnace which has a hole to permit a jet of atoms to escape. This beam is collimated by slits, and is then passed into an inhomogeneous magnetic field. A body with a magnetic moment experiences in such a field not only torque, turning the

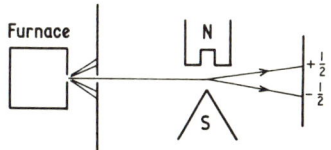

Figure 5.11   Stern—Gerlach
experiment   for   atoms   with
$J = \frac{1}{2}$

moment into line with the field, but also a deflecting force due to the
difference in field strength at the two poles of the body. The body is
therefore forced in the direction of increasing or decreasing field
strength, depending on its orientation. If the metal employed in the
experiment is sodium, it is found that the beam is split into two beams.
The ground state of Na is $^2S_{\frac{1}{2}}$ and the two beams correspond to the
two possible values of M_j, $\pm \frac{1}{2}$. This is shown schematically in
Figure 5.11. In general, if the total angular momentum of the ground
state of the atom is J, then the beam is split into $(2J + 1)$ separate
beams. This experiment is a direct demonstration that an atom may not
adopt any orientation to a magnetic field, but that only $(2J + 1)$
possible orientations are allowed.

L Molecular Beams

The Stern-Gerlach experiment is not merely of historical importance.
An almost identical arrangement provides one of the most fruitful areas
of current fundamental chemical research. The chief difference is that
the collimated beam consists not of atoms but rather of molecules.
The beam may be deflected by electric or magnetic fields and split
into separate tracks according to the value of the molecular rota-
tional angular momentum quantum number. Particular quantized sets
of molecules may be selected by the use of slits, and a variety of
experiments performed on particular quantized states. The experiments
may be spectroscopic (the investigations of transitions between one
state and another), study energy transfer (the efficacy of third bodies in
promoting energy transfer) or may even be kinetic (studying fine details
of chemical reactions). A schematic representation of the molecular
beam experiment is given in Figure 5.12.

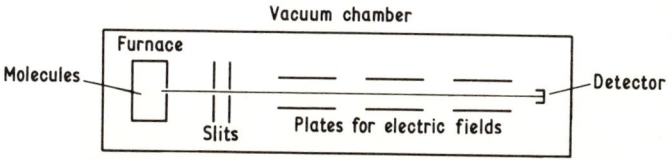

Figure 5.12   A molecular beam experiment

VI

Hyperfine Structure

When the structure of spectral lines is observed under very high resolution, it is found that in a number of cases there are more lines than can be accounted for even when Zeeman splitting has been taken into account. This extremely small splitting is called hyperfine structure, and arises from the properties of the atomic nucleus rather than those of the electrons. Two nuclear properties may give rise to hyperfine structure: isotopic mass differences and nuclear spin.

A. The Isotope Effect

The simplest atomic spectrum is that of the hydrogen atom. We have already seen in Chapter 1 that the frequencies of the lines in the hydrogen spectrum are given by:

$$h\nu = R \left\{ \frac{1}{n_1^2} - \frac{1}{n_2^2} \right\}$$

where

$$R = \frac{me^4}{8\epsilon_0^2 h^2} \tag{6.1}$$

But in reality the electron revolves not around the nucleus, but about the centre of mass of the atom; this can be accounted for by replacing m, the electronic mass, by μ, the reduced mass of the atom.

$$\mu = \frac{mM}{m + M} \tag{6.2}$$

where M is the mass of the nucleus. It is therefore clear that the observed frequencies in the hydrogen spectrum depend on the magnitude of the nuclear mass; thus the spectra of deuterium (for which the mass $M = 2$) and tritium ($M = 3$) will not be identical with that of the more common ^{1}H atom ($M = 1$). Figure 6.1 compares the Balmer series for hydrogen and deuterium. In Figure 6.1 we have considered the isotopes of hydrogen separately; in practice they occur as a mixture, and so the spectra of all three are superimposed with intensities proportional to the corresponding abundances. In naturally occurring hydrogen, the deuterium isotope has an abundance of about

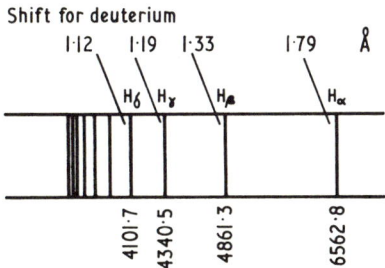

Figure 6.1   Balmer series for hydrogen showing isotopic hyperfine shifts
for deuterium

one part in 5,000, and the tritium isotope is present in negligible amounts; the lines due to the deuterium isotope have been observed, and are found at exactly the frequencies predicted by equations (6.1) and (6.2).

This sort of quantitative argument cannot be extended to more complicated atoms, as there is no analytical formula to describe the positions of spectral lines as accurately as in the H atom. Nevertheless it has been found that in many cases it is possible to represent term values, T, by empirical formulae of the type

$$T(n) = -\frac{R}{(n - \delta)^2}$$

δ is called the quantum defect and reflects the fact that in an atom such as sodium an outer electron does not experience the full nuclear charge, ze, but is screened by the inner electrons. The significance of these formulae is that the constant R still depends on the reduced mass μ, and spectral shifts still occur from one isotope to another. This effect is large enough to account for the observed spectral splittings in light atoms, but would be expected to become smaller in heavier nuclei, as the motion of the nucleus becomes less significant. This is not observed experimentally and in heavier nuclei the spectral-line shifts are at least in part due to the change of the radius of the nucleus from one isotope to another.

The existences of a number of rare isotopes have been discovered by observation of the hyperfine shifts; for example, ^{17}O was first discovered by observation of its atomic spectrum.

B. Nuclear Spin

Hyperfine structure is sometimes observable even in cases where it is certain that only a single isotopic species is present; in other cases the number of hyperfine components is greater than the number of isotopes, or may vary between different spectral lines from one element. In such circumstances the hyperfine structure is caused by the spin of the nucleus.

Nuclei are composed of protons and neutrons, and have a shell structure which is not dissimilar to the shell structure of electrons. If there are equal numbers of protons and neutrons, then the nucleus has no resultant spin; common examples are ^{12}C and ^{16}O. But if the numbers of protons and neutrons are not equal, then the resultant spin is not necessarily zero; it has a value which is an integral or half-integral multiple of $h/2\pi$, represented by $Ih/2\pi$, where I is the nuclear spin quantum number. More accurately, the spin has the magnitude $\sqrt{I(I+1)}h/2\pi$. As in other cases, we shall use the simpler form $Ih/2\pi$.

Precisely in the way that a spinning charged electron has a magnetic moment, so also has a spinning charged nucleus. The classical value of the magnetic moment μ is given by

$$\mu = -\frac{e \cdot p}{2M}$$

where $p = Ih/2\pi$. Analogous to the situation in electrons, this formula does not hold exactly, but can be corrected by the inclusion of a nuclear g-factor, so that

$$\mu = -\frac{gehI}{4\pi M}$$

As the g-factor always has a value of the order of 1, the magnetic moments of nuclei are always about 1800 times smaller than those of electrons, due to the inverse proportionality of the magnetic moment and the mass.

The nuclear angular momentum, I, and the total electronic angular momentum, J, couple together to give a resultant total angular momentum, which is designated F.

$$J + I \rightarrow F$$

The total angular momentum is quantized in units of $h/2\pi$, and F can take only integral or half-integral values. Due to this coupling, energy differences exist between states with the same values of J and I, but different values of F. However, as the magnetic moment of the nucleus is very small compared with that of the electron, these energy differences are also very small.

Figure 6.2 shows the effect of nuclear spin on a $^2P_{3/2} - {}^2S_{1/2}$ transition in the hydrogen atom. The hydrogen nucleus has $I = \frac{1}{2}$, and therefore for the lower state, where $J = \frac{1}{2}$, $F = 0$ or 1. For the upper state, $J = \frac{3}{2}$, and therefore $F = 1$ or 2. The selection rule for transitions

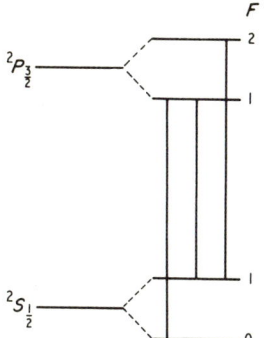

Figure 6.2 Effect of nuclear spin on a $^{2}P_{3/2} - {}^{2}S_{1/2}$ transition in hydrogen (not to scale)

to be allowed is

$$\Delta F = 0, \pm 1$$

and therefore three separate lines can be observed.

C. Hyperfine Structure in a Magnetic Field

When an atom is placed in a magnetic field, space quantization of the total angular momentum F takes place, just as we saw in the last chapter for the electronic angular momentum, J. The quantum number M_F defines the component of F in the direction of the field, and may take the values

$$M_F = -F, -F + 1, \ldots F - 1, F$$

The effect is analogous to the anomalous Zeeman effect and each state splits into $(2F + 1)$ equally-spaced components. Figure 6.3 shows the effect of a weak magnetic field on the $^{2}S_{1/2}$ ground state of the H atom. The selection rule $\Delta M_F = 0, \pm 1$ holds, and each component of a hypermultiplet has a Zeeman splitting analogous to those discussed in Chapter 5.

However, much more important is that at high field strengths a Paschen-Bach effect takes place, just as in the anomalous Zeeman

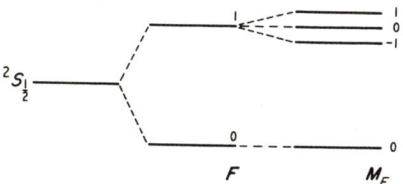

Figure 6.3 Effect of a weak magnetic field on the $^{2}S_{1/2}$ ground state of hydrogen

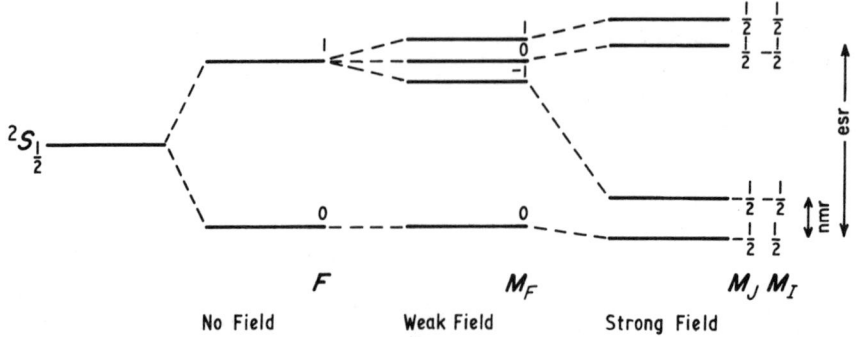

Figure 6.4   Full effects of a magnetic field on the energy levels of hydrogen $^2S_{1/2}$

effect. I and J uncouple, and M_I and M_J become individually space quantized. The only difference from the Paschen-Bach effect for ordinary multiplet structure discussed in the last chapter is that because the nuclear magnetic moment is very small, the coupling of I and J is weak, and so the Paschen-Bach effects occurs at much lower field-strengths. Figure 6.4 illustrates the full effect of a magnetic field on the $^2S_{1/2}$ ground state of the H atom. The magnetic field therefore causes the space quantization of J which we have discussed before; each term with a given value of M_J is split into $(2I + 1)$ components, and the splitting due to the nuclear spin is small compared with the splitting between terms with different values of M_J.

In the rest of this chapter we shall look at the ways in which hyperfine structure may be observed in two other important spectroscopic techniques.

D. Nuclear Magnetic Resonance Spectroscopy

In Figure 6.4, the two lowest hyperfine levels in the strong field case have the same values of M_J but different values of M_I. In the nuclear magnetic resonance experiment (n.m.r.) transitions are observed between two such levels; physically this corresponds to reversing the orientation of the nuclear spin in the magnetic field. The separation between the levels is proportional to the strength of the external magnetic field; in n.m.r. the field strengths typically obtainable in the laboratory give a separation between the levels corresponding to the energy of photons in the radiofrequency region of the spectrum. For hydrogen atoms, the frequency turns out to be about 60 MHz. The n.m.r. experiment therefore consists of placing a sample in a strong mangetic field and irradiating with radiofrequency radiation. For technical reasons it is usual to hold the radiation frequency constant and to vary the strength of the magnetic field, to find the exact field required for absorption of radiofrequency radiation, or 'resonance'.

The chemically exciting aspect of this experiment is that the magnetic field experienced by a proton in a molecule is the result not only of the external applied magnetic field, but also of magnetic fields caused by induced motions of the electrons in the molecule. If a molecule contains protons in two chemically different environments, then they will experience slightly different magnetic fields, as the induced magnetic fields will be different. Resonance will therefore take place at slightly different radiofrequencies (actually, at slightly different applied magnetic fields, as it is the field strength which is varied), and if we consider an example such as CH_3CHO, we find the spectrum has the form shown schematically in Figure 6.5.

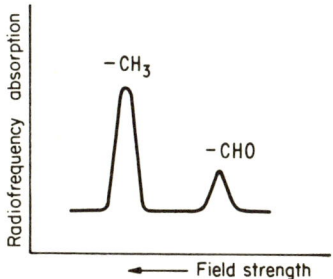

Figure 6.5   Low-resolution
n.m.r. spectrum of CH_3CHO

This shift of field required to produce resonance for each type of proton is called the chemical shift. The three $-CH_3$ protons have identical chemical environments, and so all absorb at the same energy. The $-CHO$ proton is in a different chemical environment, and absorbs the radiation at a slightly different applied field. The intensity of the $-CH_3$ peak is therefore three times the intensity of the $-CHO$ peak.

Further magnetic effects can be caused by other nuclei which have spin in the molecule, and these give rise to a hyperfine structure. Each nucleus experiences:

(a) a magnetic field due to the external field
(b) an induced magnetic field due to electronic motions (chemical shift)
(c) a magnetic field due to the possible orientations of the other spinning nuclei.

The protons in the $-CH_3$ group can experience two possible fields due to the magnetic moment of the $-CHO$ proton, as m_I can take the values $\pm\frac{1}{2}$, and so under higher resolution the $-CH_3$ peak in the

CH_3CHO spectrum is split into two peaks, with equal intensities. [Remember that C and O nuclei have no magnetic moment]. The —CHO proton can experience four different fields, depending on the values of m_I of each of the —CH_3 protons.

m_I	m_I	m_I	M_I (total)
$+\frac{1}{2}$	$+\frac{1}{2}$	$+\frac{1}{2}$	$+\frac{3}{2}$
$+\frac{1}{2}$	$+\frac{1}{2}$	$-\frac{1}{2}$	$+\frac{1}{2}$
$+\frac{1}{2}$	$-\frac{1}{2}$	$+\frac{1}{2}$	$+\frac{1}{2}$
$-\frac{1}{2}$	$+\frac{1}{2}$	$+\frac{1}{2}$	$+\frac{1}{2}$
$-\frac{1}{2}$	$-\frac{1}{2}$	$+\frac{1}{2}$	$-\frac{1}{2}$
$-\frac{1}{2}$	$+\frac{1}{2}$	$-\frac{1}{2}$	$-\frac{1}{2}$
$+\frac{1}{2}$	$-\frac{1}{2}$	$-\frac{1}{2}$	$-\frac{1}{2}$
$-\frac{1}{2}$	$-\frac{1}{2}$	$-\frac{1}{2}$	$-\frac{3}{2}$

The —CHO peak is therefore split into four peaks, with intensities in the ratio 1:3:3:1. The n.m.r. spectrum of CH_3CHO under high resolution is shown schematically in Figure 6.6. (In practice n.m.r. lines for liquids are extremely sharp).

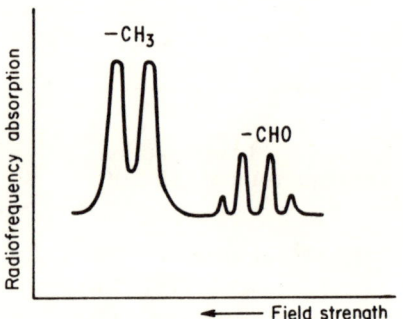

Figure 6.6   High-resolution   n.m.r. spectrum of CH_3CHO

This section gives only the briefest introduction to n.m.r., but the power of the technique is clear. The important point is that the energy levels involved are precisely those that give rise to hyperfine structure in atomic spectra.

E. Electron Spin Resonance Spectroscopy

In electron spin resonance spectroscopy (e.s.r.) transitions are observed between levels with different values of M_J; that is, the orientation of the magnetic moment of the electron is altered. With convenient laboratory produced magnetic fields, transitions between states with

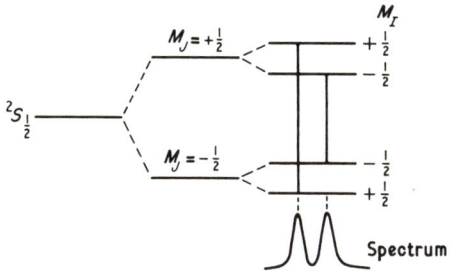

Figure 6.7   Nuclear hyperfine structure in
e.s.r.

different values of M_J correspond to microwave quanta. Figure 6.7, which is derived from Figure 6.4, shows how nuclear hyperfine structure may be observed in e.s.r. spectra. An e.s.r. spectrum can be observed for any free radical which has an unpaired electron by irradiating a sample with microwave radiation and varying an applied magnetic field to obtain absorption of the microwave radiation. In a radical, the various atomic nuclei which have nuclear spin produce hyperfine splittings of a magnitude related to the proportion of its time that the unpaired electron spends close to the particular nucleus. A simple example is that of the allyl radical (Figure 6.8) where the e.s.r. spectrum makes it clear that the unpaired electron spends nearly all its time at the two ends of the molecule, and is very rarely found close to the central carbon atom.

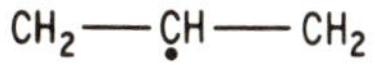

$$CH_2 —\!\!— \overset{\bullet}{C}H —\!\!— CH_2$$

Figure 6.8   The allyl radical

As with n.m.r., it is not possible here to give more than a simple outline of the technique and show how the nuclear spin gives rise to a hyperfine structure, just as in atomic spectra, and how this hyperfine structure can provide information of great importance to chemists.

VII

Experimental Aspects

Atomic spectroscopy involves the transition of atoms between electronic energy levels. Consequently the first thing we must investigate is the actual populations of atoms in the various allowed levels.

A. The Population of Atomic Energy Levels

The number of atoms, n_i, in a particular (non-degenerate) atomic energy level labelled i is given by the Boltzmann distribution law

$$n_i = n_0 e^{-\epsilon_i/kT} \qquad (7.1)$$

Here n_0 is the number of atoms in the lowest level (the ground state); ϵ_i is the excitation energy to level i; k is Boltzmann's constant and T the temperature in degrees Kelvin. Thus the population decreases exponentially as the excitation energy increases as shown in Figure 7.1.

The excitation energy of the first excited state of most atoms gives rise to transitions which appear in the visible part of the spectrum. The population of even this lowest excited state is vanishingly small at room temperature. If we consider a transition at 5,000Å, this corresponds to an excitation energy of 20,000 cm^{-1}.

$$\epsilon_i = 20,000 \text{ cm}^{-1}$$

$$= 20,000 \times 1.98 \times 10^{-23} \text{ J}$$

$$= 3.96 \times 10^{-19} \text{ J}$$

$$k = 1.38 \times 10^{-23} \text{ JK}^{-1}$$

Therefore at a temperature of 300K

$$\epsilon_i/kT = \frac{3.96 \times 10^{-19}}{1.38 \times 10^{-23} \times 300}$$

$$= 95$$

$$n_i = n_0 e^{-95}$$

$$n_i/n_0 = 5 \times 10^{-42}.$$

Thus at room temperature thermal population of the excited level may be neglected; even at 3,000K, n_i/n_0 is only 7×10^{-5}. This means that

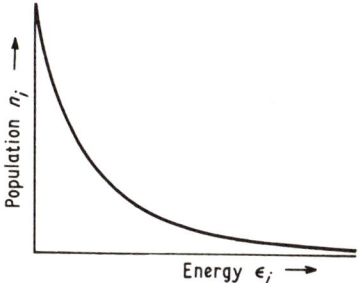

Figure 7.1   The Boltzmann distribution law

our studies of atomic spectra will be in absorption, unless energy is supplied to populate excited states from which emission is possible.

B. Atomic Absorption Spectra

Only the rare gases He, Ne, Ar, Kr and Xe actually exist as atoms at room temperature. Many other pure elements can be atomized readily if they are heated in a furnace.

To allow us to look at the spectrum of an atom in absorption, the atomic vapour must be held in a tube with windows at each end to permit white light to be shined in and passed to a spectrograph or spectrometer which produces the spectrum. This is shown in Figure 7.2. An important practical consideration is the choice of a source of continuous radiation. A simple tungsten band lamp will produce white light in the visible part of the spectrum, and in the ultraviolet region lamps containing deuterium have been used. The most powerful and convenient source which is currently employed is itself an atomic spectrum, the xenon lamp.

In the case of xenon the ionization continuum lies so low in energy that transitions to the ground state from the continuum provide an excellent white source. The actual lamps contain xenon at very high pressures, which broadens the discrete part of the spectrum (discussed later). An exciting d.c. discharge is passed through the gas.

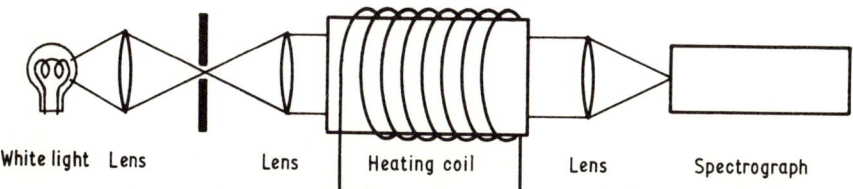

White light   Lens   Lens   Heating coil   Lens   Spectrograph

Figure 7.2   Atomic absorption spectroscopy

Furnaces can be produced simply by surrounding the sample tube (which may be metallic) with windings of resistance wire through which a current is passed, exactly as in an electric fire element. Alternatively a large current may be passed through a hollow carbon tube containing the sample.

C. Atomic Emission Spectra

Since at room temperatures most atoms will be in their ground states, the first prerequisite for emission spectroscopy is to excite the atoms to upper electronic levels.

The most obvious method of doing this is to heat the atoms, but the distribution law shows that this is unlikely to have much effect unless the excited states are extremely low-lying.

Another way of raising the temperature of atoms is to introduce them into a flame. The most obvious manifestation of this is to sprinkle salt or saline solution into a bunsen flame, when the distinctive yellow of the sodium spectrum can be seen. This is the basis of the well-known flame test for metals. Even poking a glass rod which contains sodium or sodium silicate into this flame is enough to bring out this very strong *D*-line transition. However, very few atoms are excited so that only transitions which are intrinsically very strong can be seen.

Heating a gas increases the temperature by raising the kinetic energy causing the atoms to collide. Collisions can, if energetic enough, raise the electrons to higher levels. This can be done in a more effective way if an electric discharge is passed through the gas. In this case we may even ionize the atoms and create very high effective temperatures with a consequent increase in population of excited states.

Discharges have the disadvantage that they are not particularly 'clean'. This means that as well as exciting atoms of the gas, we may produce spectral lines due to ions or atoms which derive from the materials of the electrodes.

A relatively clean method of raising atoms to excited states is by means of microwaves or radiofrequency (r.f.) discharges. The sample is put in a resonant microwave or r.f. cavity. Although individual quanta can do no more than increase the kinetic energy of the atoms, so much energy may be dissipated that the resulting excitation can be very extensive.

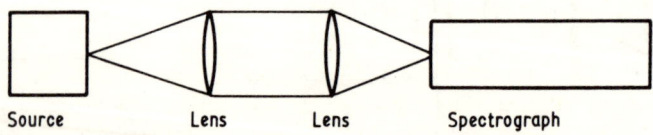

Figure 7.3   Atomic emission spectroscopy

The radiation emitted is then split into a spectrum and recorded either with a spectrograph or a spectrometer depending on the particular application, as in Figure 7.3.

D. Spectrographs and Spectrometers

The distinction between spectrographs and spectrometers is that the former are designed to measure the position or frequency of a spectral line whereas the latter measure the intensity as the most important feature.

In a spectrograph the light which is to be split into its spectrum can be dispersed either by a prism or a diffraction grating (Figure 7.4). Prisms work best at short wavelengths and are very convenient for an initial look at a spectrum since they do not have the problem found with diffraction gratings where several orders of the spectrum may overlap.

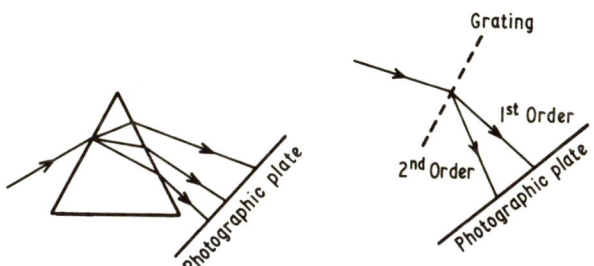

Figure 7.4   Prism and diffraction grating as dispersion units

For high resolution work a grating spectrograph is best, possibly using a prism as a predispersing unit, allowing only the wavelengths of interest to enter the spectrograph. The most popular of the modern designs of spectrograph is that of Ebert which has the diffraction grating ruled on a plane mirror (Figure 7.5). In spectrographs the dispersed spectrum is recorded on a photographic plate. The positions of lines are compared with a superimposed standard spectrum which provides wavelength standards. A correction has to be made for the refractive index of air in the spectrograph giving vacuum wavelengths. The whole process can conveniently be automated and treated by a computer.

It is clearly vital that the wavelength standards, which are normally the atomic lines of iron or thorium, are known to the greatest possible accuracy. These standard wavelengths are published from accurate measurements using a Fabry-Pérot type of interferometer (see below).

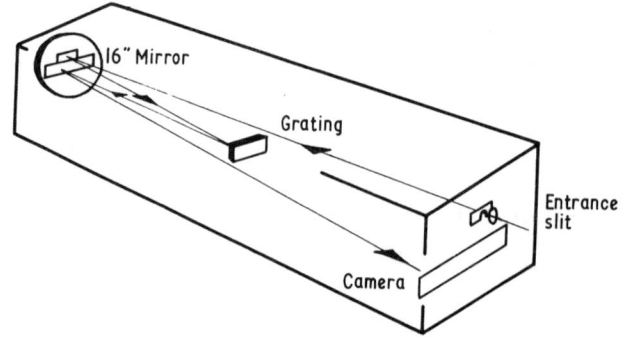

Figure 7.5   The layout of an Ebert spectrograph

In a spectrometer the spectrum is recorded not photographically but by a photomultiplier. In this case the photoelectric current is directly proportional to the number of photons striking the detector, and thus the measurements are useful when intensity comparisons are required.

E. Interferometry

Accurate values of wavelength standards are determined by interferometry usually using a Fabry-Pérot interferometer (Figure 7.6). A Fabry-Pérot etalon comprises two plane parallel glass plates which part reflect and part transmit incident light. A telescope can be used to observe a set of circular rings caused by interference of light rays which have suffered differing numbers of traversals of the space between the plates due to multiple reflections. The condition for a maximum to occur can be shown to depend on the wavelength of the light and the separation of the parallel plates; a sharp maximum in intensity is reached whenever the optical path between the mirrors is equal to an integral number of half-wavelengths of the incident radiation.

In practice the optical path length between the plates can be altered by adjusting the air pressure within the etalon enclosure. By counting

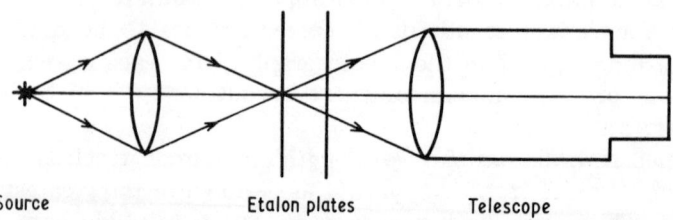

Source   Etalon plates   Telescope

Figure 7.6   The Fabry-Pérot interferometer

the number of fringes observed for a given path-length change, a wavelength can be determined.

Ultimately all length measurements must be related to a standard length. This is no longer a bar of metal of length one metre, but an actual atomic transition wavelength. Other secondary standards may be determined by comparison of wavelength coincidences using a Fabry-Pérot etalon rather as pendulum periods may be compared by observing the interval between coincidences.

F. Intensities

The intensity of a spectral line depends on three factors: the population of the initial energy level, statistical weight effects and the intrinsic probability of the transition.

The initial population depends on the Boltzmann factor, that is on the temperature and the excitation energy of the initial state. This was given in equation (7.1). However, if we look at the sodium D lines, the intensities are (as we saw in Chapter 5) in the ratio 2:1, due to the relative statistical weights of the $^2P_{3/2}$ and $^2P_{1/2}$ levels being 4 and 2. We can include this factor by writing the Boltzmann law in its more general form

$$n_i = n_0 g_i e^{-\epsilon_i / kT}$$

where g_i is the statistical weight of the level i.

The intrinsic probability of the transition, I, is given by

$$I = \int \Psi_I R \Psi_{II} d\tau,$$

as we saw when discussing selection rules. Ψ_I and Ψ_{II} are the wave functions of the initial and final states, and R is the operator for the commonly observed electric dipole transitions. We have seen how symmetry arguments may be used to predict when I is non-zero.

Practically, the intensity of a transition in absorption is measured by the absorption coefficient, ϵ. This is defined by the Beer—Lambert law

$$I/I_0 = e^{-\epsilon cl}$$

where I and I_0 are the light intensities with and without the sample, c is the concentration of the vapour and l is the path-length.

G. The Temperature of a Flame

The temperature of a flame (Figure 7.7) may be measured by volatilizing sodium chloride in the flame and observing the yellow sodium D lines, $^2P \rightarrow {}^2S$. Comparison is made between the brightness of the sodium lines excited in the flame and the brightness (at the same wavelength) of a tungsten strip filament lamp whose temperature can

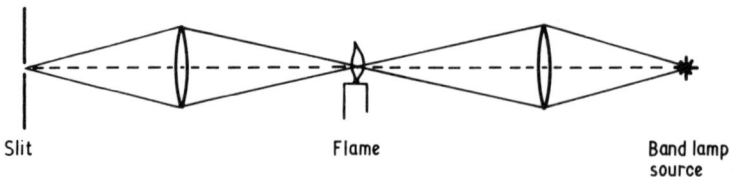

Slit   Flame   Band lamp
                                                                  source

Figure 7.7   The temperature of a flame

be controlled by a variable transformer. The temperature of the
filament lamp is then measured by an optical pyrometer.

When the brightness temperature of the lamp is lower than that of
the sodium in the flame, the sodium lines will appear in emission;
conversely, when the temperature of the lamp is higher, the sodium
lines will be reversed, i.e. will appear in absorption.

The intensity of an emission line is usually determined by the
concentration of the emitting state — itself determined, in equilibrium,
by the concentration of unexcited atoms and the temperature — and by
the probability of emission. However, in the present case we make the
assumption that the brightness of the flame is determined just by the
temperature. This assumption is made possible because the $^2P \rightarrow {}^2S$
transition is a resonance transition. Most atoms in the flame are in the
ground, ^2S, state, so that emission of a photon is almost always
followed by reabsorption. The reabsorption excites the ^2S atom to the
^2P state which then emits a photon which is absorbed yet again. The
photons are said to be 'trapped', and the yellow radiation escapes only
from a surface layer around the flame. Increasing the concentration of
Na increases the density of emitting species but decreases the thickness
of the emitting layer, and the intensity remains constant. The flame
containing the sodium atoms behaves as a near-perfect emitter and
absorber, but only at the wavelength of the sodium resonance
transition. That is, the flame is a 'perfect yellow body'. The arguments
used to develop the black body 'E_λ versus λ' curve now show that the
intensity of the flame at the yellow lines is identical to the intensity at
that wavelength of a black body at the same temperature.

H. Linewidths

So far when discussing spectral lines we have not mentioned any actual
width for the lines. In fact they are not infinitely sharp but have a
profile such as that illustrated in Figure 7.8.

There are three reasons why the lines are not infinitely sharp. The
first is that each line has a so-called natural linewidth. This results from
the Uncertainty Principle. In the most commonly encountered form

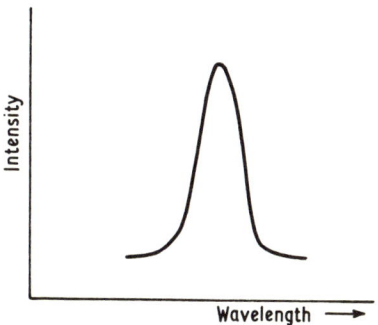

Figure 7.8 The intensity distribution of a spectral line

Heisenberg's principle is written as

$$\Delta p \cdot \Delta x \approx h/4\pi$$

where Δp is the uncertainty in momentum and Δx the uncertainty in position. This relation can be recast in terms of energy and time instead of momentum and position

$$\Delta E \cdot \Delta t \approx h/4\pi$$

In this form we gain some insight into the natural width, ΔE, of a spectral line; it is inversely proportional to the lifetime of an atom in the particular state. Thus if an absorption is from a long-lived ground state of an atom up to a short-lived excited state, then the line is broadened because of the small value of Δt for the upper state.

A second mechanism which gives rise to broadening of spectral lines is the Doppler effect. Just as the sound waves from a moving object appear to be higher frequency if the source is moving towards the observer and lower if it is moving away, the same principle applies to light waves being emitted by an atom. Since in most spectroscopic experiments some atoms will be moving towards the spectrograph and some away depending on a Boltzmann distribution of kinetic energies, then the lines observed are not infinitely sharp but broadened. This effect will become greater with increasing temperature, as a wider range of atomic velocity becomes possible at elevated temperatures.

The third contributor to line-broadening is the effect of pressure. As pressure increases the atoms in a gas are increasingly involved in collisions. The photons may then be emitted from a pair of atoms which happen to be close together at the moment of emission. It is a fundamental tenet of quantum mechanics that when two identical systems come close together the wave function which represents the pair is either the sum or difference of the individual wave functions and the two corresponding energies are one greater and one less than the

84

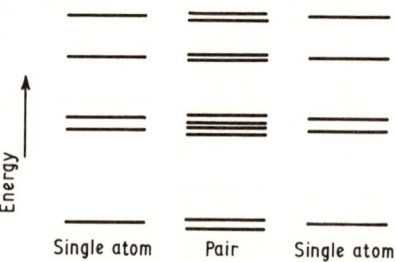

Single atom   Pair   Single atom

Figure 7.9 Resulting energy levels of two atoms forming a loosely bound pair

separate individuals. This is shown schematically in Figure 7.9. The closer together the individual atoms come the wider the splitting. In a dense gas a whole range of separations of pairs are to be found at any one time with the result that the actual energy levels will appear more like Figure 7.10. The bigger the pressure the more pronounced is the effect.

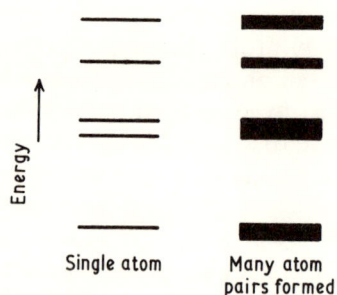

Single atom   Many atom
                   pairs formed

Figure 7.10 The effect of pressure broadening

I. Lasers

Extremely narrow spectral lines can be obtained if we use laser action. So far we have considered two processes involving the interaction of matter with photons, absorption and emission. To be strictly accurate we ought to be more precise and call what we have so far been concerned with, stimulated absorption and spontaneous emission. There is a third possibility—stimulated emission as illustrated by the three diagrams below:

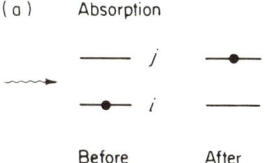

(a)   Absorption

Before   After

The electron is excited to the upper level. Number of transitions = $BN_i\rho(\nu)$, where N_i is the number of atoms in the lower state, $\rho(\nu)$ the density of radiation and the constant, B, is called the Einstein B coefficient

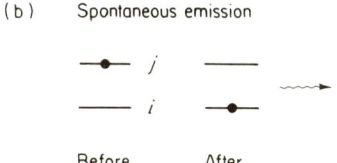

(b)   Spontaneous emission

Before   After

The electron falls from the upper level and a photon frequency ν is emitted; the number of photons being AN_j

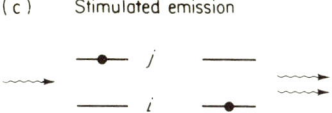

(c)   Stimulated emission

Number of transitions is $BN_j\rho(\nu)$.

In stimulated emission the incoming photon is augmented so that the two photons emitted are exactly in phase.

For stimulated emission to be the major effect we must have N_j greater than N_i which cannot happen at equilibrium, following the Boltzmann distribution law. The condition where there is a greater population in an upper level than a lower level may be achieved by using a photographic flash on a ruby crystal containing Cr^{3+} or Eu^{2+} ions. The excited crystal loses its energy to the ions so that momentarily we do have $N_j > N_i$. Alternatively, atoms such as neon atoms can be continuously given energy by collision with excited helium atoms, thus maintaining a population in an excited level which can have transitions to a previously unoccupied lower (but still excited) level.

This stimulated emission of radiation (SER) becomes a LASER by light amplification (LA). Amplification is achieved by having the atoms in a box like a Fabry-Pérot interferometer and allowing the light to bounce back and forth perpendicular to the end mirror which only allows a fraction of the light to be emitted.

The emitted light has unique properties. It has a very narrow linewidth, it is directional and it is coherent. These amazing properties have led to an abundance of applications from the purely scientific to some which almost overlap with science fiction. It is salutary to remember that most of the light which comes from lasers is in the form of lines from atomic spectra. Atomic spectroscopy was in at the birth of quantum mechanics and yet it remains a powerful and fascinating subject throwing light on to the dark areas of theory and providing tools for the most elegant of experiments.

VIII

Applications of Atomic Spectroscopy

A. Atomic Ground States and Chemistry

The interpretation of the chemistry of the elements in terms of their electron configurations is now so familiar to us that we rarely stop to wonder how the electron configurations were originally obtained. In this section we shall show what atomic states are obtained from some simple electron configurations, and hence show how the electron configuration of an element may be deduced from the experimentally determined ground state term type.

A simple example is the potassium atom; this has the electron configuration

$$1s^2 \ 2s^2 \ 2p^6 \ 3s^2 \ 3p^6 \ 4s$$

and the presence of a single $4s$ electron outside the closed shell core means that the ground state is ^2S. The fact that the experimental ground state is found to be ^2S allows us to discard another possible electron configuration for potassium,

$$1s^2 \ 2s^2 \ 2p^6 \ 3s^2 \ 3p^6 \ 3d$$

This would give a ^2D ground state; the splitting of lines obtained in a magnetic field clearly shows that the K ground state is ^2S and not ^2D.

A less familiar example is found in the case of Cr. Table 8.1 shows the configurations of some elements in the first transition series. The configurations are of importance because they lead directly to the valencies, and of particular interest is the configuration of Cr, which is out of step with its neighbours. The evidence for this assignment is that the ground state of Cr is found to be ^7S. We can readily write down the quantum numbers of the various electrons if the configuration is $4s^1 \ 3d^5$:

n	l	m_l	m_s	
4	0	0	½	($4s$)
3	2	2	½	($3d$)
3	2	1	½	
3	2	0	½	
3	2	−1	½	
3	2	−2	½	
		0	3	

Table 8.1

Sc	$4s^2$	$3d^1$
Ti	$4s^2$	$3d^2$
V	$4s^2$	$3d^3$
Cr	$4s^1$	$3d^5$
Mn	$4s^2$	$3d^5$
Fe	$4s^2$	$3d^6$

As required, the total orbital angular momentum is zero, and the total spin angular momentum is 3. However, a ^7S state cannot arise from the configuration $4s^2\ 3d^4$; the two $4s$ electrons must have opposite spins, by the Pauli principle, and there is no way that four $3d$ electrons could give rise to a spin angular momentum of 3, which requires at least 6 unpaired electrons.

B. Excited Electronic Energy Levels

Although a knowledge of the ground state electronic configurations of atoms is of paramount importance, it is not sufficient for an understanding of chemistry. The ground state properties are those of the unperturbed isolated atom; the chemistry is related to energy changes on reaction, when the atom is influenced by other atoms, and electrons are shared or transferred. Thus in addition to ground state information we ideally would like to know as much as possible about how easy it is to involve the outer valence electrons in chemical change.

An elementary example of this sort of consideration is found in the chemistry of carbon. Carbon has the ground state configuration

$$1s^2\ 2s^2\ 2p^2$$

in which there are two unpaired electrons. Yet by far the most common form of bonding which carbon undergoes is the formation of four equivalent single covalent bonds, rather than the two covalent bonds which might be expected from using the $2p$ electrons in bonding. This is often explained in the following way. One of the $2p$ electrons is promoted to an unfilled $2p$ orbital, giving four unpaired electrons. These undergo 'hybridization', by which four equivalent orbitals, designated sp^3, are formed, each containing one electron. This type of argument must rest heavily on the energy obtained from the formation of two extra bonds exceeding the promotion energy of the $2s$ electron; an estimate of the promotion energy of the $2s$ electron may be obtained directly from an interpretation of the excited electronic states of the carbon atom.

C. Ionization Potentials

The ease with which an electron can be removed completely from an atom, rather than just excited to a higher level, is given quantitatively

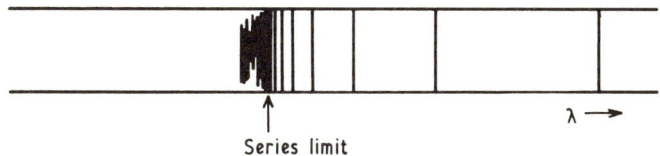

Figure 8.1 Ionization potential from a series limit

by the ionization potential. In terms of the observed spectra, the ionization potentials correspond to series limits. In many cases these series limits may be difficult to measure accurately and appear as a series of lines gradually merging into a region of continuous absorption, as in Figure 8.1. If the spectrum cannot be observed right up to the series limit, it is often possible to extrapolate to this limit. The energy levels involved in such a series may often by represented by the formula

$$T(n) = R/(n - \delta)^2$$

Here δ is an empirical correction called the quantum defect. In a polyelectronic atom the various types of outer electron (s,p,d, etc) penetrate the inner shells to different extents and hence experience differing apparent nuclear charges. Table 8.2 shows the δ values for the states of sodium which arise from electron configurations

$$1s^2\, 2s^2\, 2p^6\, (nl)$$

Thus the quantum defect is in the order

$$s > p > d$$

and is negligible for f and g electrons. Alternatively, the effect may be expressed in terms of the charge seen by the outermost electron. For Na, the values are as shown in Table 8.3.

Atomic spectra may be observed not only for neutral atoms, but also for positively charged ions. Clearly the first ionization potential of a doubly charged ion is equal to the second ionization potential of the singly charged ion, and the third ionization potential of the neutral

Table 8.2 Values of δ the Quantum Defect, for Some States of Na

$l =$	0	1	2	3	4
n	s	p	d	f	g
10	1.350	0.856	0.014	0.006	—
7	1.350	0.858	0.014	0.004	—
5	1.353	0.862	0.013	0.002	0.000
4	1.357	0.865	0.012	0.001	—
3	1.357	0.883	0.010	—	—

Table 8.3   Effective Charges Seen by Outer Electrons in Na

	s	p	d	f	g
10	1.156	1.093	1.001	1.001	—
7	1.239	1.140	1.002	1.000	—
5	1.371	1.208	1.003	1.000	1.000
4	1.513	1.277	1.003	1.000	—
3	1.844	1.417	1.003	—	—

atom. Ionization potentials are now known with reasonable accuracy for most atoms; a summary of some of the results obtained from atomic spectra is given in Table 8.4. These data are essential to all calculations based on the ionic model for the prediction of the relative stabilities of compounds, and for the investigation of the nature of bonding in crystals.

Table 8.4   Ionization Potentials (electron volts)

Element	1st IP	2nd IP	3rd IP	4th IP
H	13.6			
He	24.6	54.4		
Li	5.4	75.6	122.4	
Be	9.3	18.2	153.9	217.7
B	8.3	25.1	37.9	259.3
C	11.3	24.4	47.9	64.5
N	14.5	29.6	47.6	77.4
O	13.6	35.1	55.1	77.3
F	17.4	35.0	62.6	87.1
Ne	21.6	41.0	63.4	96.9
Na	5.8	47.3	71.7	—
Mg	7.6	15.0	80.1	109.5
Al	6.0	18.8	28.4	120.0
Si	8.1	16.3	33.5	45.1
P	11.0	19.7	30.1	51.4
S	10.4	23.4	35.0	47.3
Cl	13.0	23.8	39.9	54.5
Ar	15.8	27.6	40.7	61.0

D. Magnetic Properties

We have seen in Chapter V that if an atom has a total angular momentum quantum number J, then as a spinning particle it has associated with it a magnetic moment μ given by

$$\mu = -\sqrt{J(J+1)}g\beta$$

where β is the Bohr magneton, and g depends on the values of L, S and J for the ground state. It is possible to measure the magnetic moment directly by the Stern—Gerlach experiment (see Chapter V), in which a beam of atoms is passed through an inhomogeneous magnetic field.

We can also obtain information on the magnetic moments of atoms by measuring the paramagnetism of a bulk sample, either a gas, or in some cases a solution or crystal. An atom with an angular momentum quantum number J has $(2J + 1)$ possible values for the quantum number M_J, the component of the angular momentum, and in the absence of a magnetic field these different states are all degenerate. In a bulk sample, all the states will be equally populated, and as M_J takes values from $-J$ to $+J$, the sample will have no resultant angular momentum or magnetic moment. However, if a magnetic field is applied, then states with different values of M_J will have different energies, and therefore different populations. These populations will be governed by the Boltzmann distribution law, which we discussed in the last chapter. For a given field, the differences in population between the states increase as the temperature decreases, whereas for a given temperature, the differences in population increase as the field strength increases. The atoms tend to align their magnetic moments with the field, this being the most stable arrangement, and so the bulk sample acquires a net magnetic moment; the magnetization is in the same direction as the applied field, and the sample is said to be paramagnetic.

If P is the magnetic moment per unit volume, and N is the number of atoms per unit volume, then it may be shown using the Boltzmann distribution law that

$$P = \frac{J(J + 1)g^2 \beta_0^2 N}{3kT} \cdot B$$

The coefficient of B is called the magnetic susceptibility, K.

$$\therefore \quad K = \frac{J(J + 1)g^2 \beta_0^2 N}{3kT} = \frac{\mu^2 N}{3kT}$$

This is the Curie Law, that the susceptibility is inversely proportional to the absolute temperature; the derivation holds only provided that the temperature is not too low, nor the magnetic field B too high.

An important application of paramagnetism is the technique of adiabatic demagnetization. Many of the salts of the lanthanides contain ions which have magnetic moments. When a magnetic field is applied to these salts, the magnetic moments tend to align themselves with the field; as this is a position of lower energy than before, a small amount of energy is transferred to the vibrational motions in the solid. Applying the magnetic field is therefore accompanied by a small increase in temperature, though this effect is so small that it can only be observed at low temperatures.

The converse of this process, removing the magnetic field, therefore causes a small cooling effect, and this provides the basis for adiabatic demagnetization. A sample of a rare earth salt at a low temperature is placed in good thermal contact with its surroundings. A magnetic field is applied, and the heat evolved is efficiently conducted away to the surroundings. The thermal contact is then broken, and the magnetic field removed; the temperature of the salt now drops. This procedure may be repeated, and in this way it is possible to obtain temperatures of 0.001K; applying the same method to nuclear magnetic moments has produced temperatures of 2×10^{-5} K.

E. Astrophysical Applications

In interstellar space where the density of matter is very low there exist free atoms whose radiation may be detected on earth. From the intensities of such sources is built up the knowledge we have of the abundance of the elements in space. Quantitative estimates of the percentage abundances of the elements provide the data upon which cosmological theories may be tested. One example is the ratio of hydrogen to deuterium, which is a critical factor in dating the universe if a 'big-bang' theory is assumed. Although hydrogen may be formed in a number of atomic processes, deuterium could only be formed in the short time following the initial bang.

The elements which are to be found in the atmosphere of stars, such as our own sun, are also known to us from studies of atomic spectra. The sun emits white light, that is spectrally continuous light, and some of this light is absorbed by atoms in its outer atmosphere. This type of process gives rise to the well-known Fraunhofer absorption lines in the spectrum of sunlight. By comparing the frequencies of lines in stellar spectra with those obtained in laboratory spectra, and assuming that the laws of physics are identical on earth and in distant stars, the nature of the elements may be determined, and from intensity measurements it is also possible to make estimates of relative abundances.

When the absorption or emission spectra of stars very remote from earth are observed, it is found that the atomic line spectra are shifted slightly towards the red end of the spectrum. This is interpreted as being due to an optical Doppler effect which we met in the last chapter in connection with line broadening. The Doppler effect is the change in frequency of a signal observed by a stationary observer due to the relative motion of the source. From the astrophysical spectral shifts it is possible to determine the velocities of the stars relative to the earth. It is on this evidence that the theory of the expanding universe is based.

A large part of the experimental data for the study of astrophysical problems comes from studies of atomic spectra. This type of research has been given new impetus with the advent of satellites, which can now be used to observe phenomena at frequencies which have hitherto been obscured by absorption or scattering caused by our own atmosphere. As a result the study of atomic spectra remains an important area of basic research, not least in the chemistry of interstellar space and the origin of the planets, molecules and even life itself.

Suggestions for Further Reading

Condon, E. U. and G. H. Shortley (1963). *The Theory of Atomic Spectra*, Cambridge University Press.
Herzberg, G. (1944). *Atomic Spectra and Atomic Structure*, Dover, New York.
Kuhn, H. (1962). *Atomic Spectra*, Longmans, London.

The fundamental constants in SI units

Avogadro constant	L or N_A	6.022×10^{23} mol^{-1}
Bohr magneton	β	9.274×10^{-24} JT^{-1}
Bohr radius	a_0	5.292×10^{-11} m
Boltzmann constant	k	1.381×10^{-23} JK^{-1}
charge of a proton	e	1.602×10^{-19} C
gas constant	R	8.314 JK^{-1} mol^{-1}
nuclear magneton	μ_N	5.051×10^{-27} JT^{-1}
permeability of a vacuum	μ_0	4×10^{-7} H m^{-1} or NA^{-2}
permittivity of a vacuum	ϵ_0	8.854×10^{-12} F m^{-1}
Planck constant	h	6.626×10^{-34} Js
(Planck constant)/2π	$\hbar$	1.055×10^{-34} Js
rest mass of electron	m_e	9.110×10^{-31} kg
rest mass of proton	m_p	1.673×10^{-27} kg
speed of light in a vacuum	c	2.998×10^8 m s^{-1}

$\ln 10 = 2.303$ $\ln x = 2.303 \lg x$ $\lg e = 0.4343$ $\pi = 3.142$
$R \ln 10 = 19.14$ JK^{-1} mol^{-1}.

Energy equivalents
  1 electron volt (eV) = 1.602×10^{-19} Joules (J)
  1 electron volt per molecule = 96.48 kJ mol^{-1}
  1 erg = 10^{-7} J
  1 wave number (cm^{-1}) = 1.986×10^{-23} J

Problems

Chapter 1

1. In a photo-electric effect experiment, using a target of copper, it is found that a current is produced only if the incident light has a wavelength less than 3105Å. What is the threshold frequency for copper? Calculate the value of the work function, expressing your answer in Joules.

2. Given the wavelengths (Å) of the lines in the Balmer series as follows, calculate a value for the Rydberg constant:

   6562.79, 4861.33, 4340.47 and 4101.74.

3. What is the velocity of an electron in the ground state of the hydrogen atom?

4. Calculate the Bohr radius (the radius of the first orbit) for He^+.

5. Give a qualitative explanation in terms of energy gaps for the colours of the halogens: fluorine — colourless, chlorine — yellow-green, bromine — red, iodine — purple. What colour would you expect the next member of the series, astatine to be?

Chapter 2

1. Calculate the de Broglie wavelength of a hydrogen molecule moving at 100 m s^{-1}. (Take the mass of a hydrogen molecule to be twice the mass of a proton)

2. Calculate the first three energy levels of a proton in a one-dimensional box of length 10^{-10} m. What are the energies of the first three levels of a proton in a three-dimensional box with sides of 10^{-10} m each?

3. Use the de Broglie relationship to calculate the wavelength of the electron in the ground state of the hydrogen atom. Compare your result with the circumference of the first Bohr orbit.

4. Use the free electron model to predict the variation in the absorption spectra of conjugated long chain hydrocarbons with chain length. Suppose that there are $2N$ carbon atoms, and that the bond length is L; write down the energy of the nth level, supposing the box extends $\frac{1}{2}L$ beyond each end of the molecule. Each atom provides one electron to the conjugated system, and each level has two electrons in it, so the highest occupied level has $n = N$. The first transition is thus from $n = N$ to $n = N + 1$. Write down an expression for the energy of this transition, and show that as the chain length increases, the energy of the first transition decreases.

Chapter 3

1. Give the electron configurations for the following elements: N, Mn, Se, W, Np.

2. How many rare-earth elements are there in a single row, and why should this be so? If a series of super-heavy elements could be produced artificially by filling the 5g orbitals, how many such elements would exist?

3. Explain how the valency of an element is related to its position in the Periodic Table. How does variable valency arise?

Chapter 4

1. What values may J take in the following terms; ^1P, ^2D, ^5D.

2. Give term symbols, in Russell—Saunders coupling, for the ground states of the following atoms: B, N, F, As.

3. How would you expect the splittings of the sub-levels of the excited ^2P states of He^+ to compare with the corresponding splittings in the case of H?

4. Why are theoretically derived selection rules not always obeyed in practice?

Chapter 5

1. Show, by means of a diagram how a ^1D level would be split by a magnetic field. How is a ^2D level split by a magnetic field?

2. What would be the relative intensities of the components of a $^3P - ^3S$ transition?

3. What is the term symbol for the ground state of the Sc atom? What pattern would be observed in the Stern—Gerlach experiment if Sc atoms were used?

Chapter 6

1. Calculate the reduced mass of H and D; hence calculate the wavelength of the line in the D spectrum which corresponds to the line in the H spectrum at 6562.79Å.

Index